I0789943

Wide-Bandgap Semiconductor Materials and Devices 11 -and- State-of-the-Art Program on Compound Semiconductors 52 (SOTAPOCS 52)

Editors:

J. Kim
Korea University
Seoul, South Korea

E. B. Stokes
University of North Carolina
Charlotte, North Carolina, USA

G. W. Hunter
NASA Glenn Research Center
Cleveland, Ohio, USA

S. Sarkozy
University of Cambridge
Cambridge, United Kingdom

Sponsoring Divisions:

 **Electronics and Photonics**

 **Sensor**

Luminescence and Display Materials

Published by
The Electrochemical Society
65 South Main Street, Building D
Pennington, NJ 08534-2839, USA
tel   609 737 1902
fax   609 737 2743
www.electrochem.org

ecstransactions ™

Vol. 28, No. 4

Copyright 2010 by The Electrochemical Society.
All rights reserved.

This book has been registered with Copyright Clearance Center.
For further information, please contact the Copyright Clearance Center,
Salem, Massachusetts.

Published by:

The Electrochemical Society
65 South Main Street
Pennington, New Jersey 08534-2839, USA

Telephone 609.737.1902
Fax 609.737.2743
e-mail: ecs@electrochem.org
Web: www.electrochem.org

ISSN 1938-6737 (online)
ISSN 1938-5862 (print)
ISSN 2151-2051 (cd-rom)

Printed in the United States of America.

Preface

The papers included in this issue of *ECS Transactions* were originally presented in the symposium "Wide-Bandgap Semiconductor Materials and Devices 11", held during the 217th meeting of The Electrochemical Society, in Vancouver, Canada from April 25-30, 2010.

***ECS Transactions*, Volume 28, Issue 4**
Wide-Bandgap Semiconductor Materials and Devices 11 -and- State-of-the-Art Program on
Compound Semiconductors 52 (SOTAPOCS 52)

Table of Contents

Preface   *iii*

Chapter 1
ZnO

Studies of Electron Trapping in ZnO Semiconductor   3
  L. Chernyak, E. Flitsyian, M. Shatkhin and Z. Dashevsky

Aluminum-Doped Zinc Oxide Thin Films for Opto-Electronic Applications   13
  N. Hirahara, B. Onwona-Agyeman and M. Nakao

Investigation and Fabrication of Bottom Gate ZnO:Al TTFTs with Various Thicknesses   21
of ZnO Buffer Layers
  Y. Lin, H. Lee and C. Lee

High Responsivity Ultraviolet Photodetector Based on p-GaN/i-ZnO Nanorod /n-ZnO: 27
In Nanorod
  C. Chen, J. Yan and C. Lee

Epitaxial Growth of ZnO on $LiAlO_2$ and $LiGaO_2$ Substrates by Chemical Vapor   33
Deposition
  C. Chen, J. Yu, T. Huang, L. Chang and M. Chou

Chapter 2
III-Nitrides

Quasi-Ballistic Hole Transport in an AlGaN/GaN Nanowire   47
  M. Mastro, H. Kim, J. Ahn, J. Kim, J. Hite and C. Eddy Jr.

Light Extraction Enhancement of n-Side Up Light-Emitting Diodes Without Electrodes   53
Covering by Wafer Bonding and Textured Surfaces
  R. Horng, Y. Lu and D. Wuu

v

AlGaN/GaN HEMT Based Biosensor   61
S. Alur, T. Gnanaprakasa, Y. Wang, Y. Sharma, J. Dai, J. Hong, A. L. Simonian,
M. Bozack, C. Ahyi and M. Park

Development of Enhancement Mode AlN/Ultrathin AlGaN/GaN HEMTs by Selective   65
Wet Etching
T. Anderson, M. Tadjer, M. Mastro, J. Hite, K. Hobart, C. Eddy Jr. and F. Kub

ZnO Nanorod/p-GaN Heterostructured Light-Emitting Diodes Passivated Using   71
Photoelectrochemical Method
J. Yan and C. Lee

Chapter 3
SiC and Related Materials

Etch Pits of 4H-Silicon Carbide Surface Formed Using Chlorine Trifluoride Gas   81
H. Habuka, K. Furukawa, K. Tanaka, Y. Katsumi, N. Takechi, K. Fukae and T. Kato

a-Plane GaN for Hydrogen Sensing Applications   89
K. Baik, W. Lim, S. Pearton, Y. Wang, F. Ren, J. Yang and S. Jang

Passivation of Deep Levels at the SiO_2/SiC Interface   95
A. F. Basile, J. Rozen, X. Chen, S. Dhar, J. R. Williams, L. C. Feldman and
P. M. Mooney

Chapter 4
General Poster Session

pnpn and npn Heterostructural Optoelectronic Devices   105
D. Guo

npn Heterostructural Optoelectronic Switch with Collector Confinement Layer   111
D. Guo

Drain Leakage Current in MuGFETs at High Temperatures   119
J. Giroldo Jr. and M. Bellodi

Fabrication of IGZO Sputtering Target and Its Applications to the Preparation of   131
Thin-Film Transistor Devices
C. C. Lo and T. Hsieh

Low-Resistivity and High-Transmittance Indium Gallium Zinc Oxide Films Prepared by Co-Sputtering $In_2Ga_2ZnO_7$ and In_2O_3 Targets   137
H. Chang, K. Huang, C. Chu, S. Chen, T. H. Huang and M. Wu

InGaN-Based Light Emitting Diodes with an AlN Sacrificial Buffer Layer for Chemical Lift-Off Process   149
C. Lin, J. Dai and M. Lin

InGaN Light-Emitting Diode Structure on a Photoelectrochemical Treated GaN:Si Layer   155
K. Chen, C. Lin and C. Lin

Electron Paramagnetic Resonance Studies of Shallow Donors Behavior in Hydrogenated ZnO Films   161
L. Larina, N. Tsvetkov, J. Yang, K. Lim and O. Shevaleevskiy

Dy^{3+} Emission from GaAlN Powder and Radio-Frequency Sputtered Thin Film   169
J. H. Tao, J. McKittrick, J. Talbot and K. C. Mishra

Chapter 5
Energy Devices

Cuprous Oxide Solution Preparation and Application to Cu_2O-ZnO Solar Cells   179
A. Du Pasquier, Z. Duan, N. Pereira and Y. Lu

Author Index   191

Facts about ECS

The Electrochemical Society (ECS) is an international, nonprofit, scientific, educational organization founded for the advancement of the theory and practice of electrochemistry, electrothermics, electronics, and allied subjects. The Society was founded in Philadelphia in 1902 and incorporated in 1930. There are currently over 7,000 scientists and engineers from more than 70 countries who hold individual membership; the Society is also supported by more than 100 corporations through Corporate Memberships.

The technical activities of the Society are carried on by Divisions. Sections of the Society have been organized in a number of cities and regions. Major international meetings of the Society are held in the spring and fall of each year. At these meetings, the Divisions and Groups hold general sessions and sponsor symposia on specialized subjects.

The Society has an active publications program that includes the following.

Journal of The Electrochemical Society — JES is the peer-reviewed leader in the field of electrochemical and solid-state science and technology. Articles are posted online as soon as they become available for publication. This archival journal is also available in a paper edition, published monthly following electronic publication.

Electrochemical and Solid-State Letters — ESL is the first and only rapid-publication electronic journal covering the same technical areas as JES. Articles are posted online as soon as they become available for publication. This peer-reviewed, archival journal is also available in a paper edition, published monthly following electronic publication. It is a joint publication of ECS and the IEEE Electron Devices Society.

Interface — *Interface* is ECS's quarterly news magazine. It provides a forum for the lively exchange of ideas and news among members of ECS and the international scientific community at large. Published online (with free access to all) and in paper, issues highlight special features on the state of electrochemical and solid-state science and technology. The paper edition is automatically sent to all ECS members.

Meeting Abstracts (formerly Extended Abstracts) — Abstracts of the technical papers presented at the spring and fall meetings of the Society are published on CD-ROM.

ECS Transactions — This online database provides access to full-text articles presented at ECS and ECS-sponsored meetings. Content is available through individual articles, or as collections of articles representing entire symposia.

Monograph Volumes — The Society sponsors the publication of hardbound monograph volumes, which provide authoritative accounts of specific topics in electrochemistry, solid-state science, and related disciplines.

For more information on these and other Society activities, visit the ECS website:

www.electrochem.org

CHAPTER 1

ZnO

2

Studies of Electron Trapping in ZnO Semiconductor

L. Chernyak[a], E. Flitsiyan[a], M. Shatkhin[a], and Z. Dashevsky[b]

[a] Department of Physics, University of Central Florida, Orlando, Florida 32816, USA
[b] Department of Materials Engineering, Ben-Gurion University, Beer-Sheva 84105, Israel

> It has been recently discovered that electron injection into Phosphorus-, Lithium-, Antimony- or Nitrogen-doped ZnO semiconductor, using electron beam from a Scanning Electron Microscope, as well as a forward bias application to the p-n junction or Schottky barrier, leads to a multiple-fold increase of minority carrier diffusion length and lifetime (1-4). It has also been demonstrated that forward biasing a ZnO-based photovoltaic detector results in a several-fold responsivity enhancement due to a longer minority carrier diffusion length in the detector's p-region as a result of electron injection (5, 6). The observed electron injection effects were attributed to the charging of the meta-stable centers associated with the above-referenced impurities.

With p-type doping of ZnO becoming possible, it is very likely that minority carrier (bipolar) devices such as LEDs, laser diodes, and transparent p-n junctions can be achieved in the near future (7, 8, 9). Besides effective p-type doping and robust ohmic contact fabrication, attaining the optimum performance of bipolar devices hinges upon overcoming an additional challenge - a short minority carrier diffusion length (usually $\leq 1\mu m$), which is common to direct band gap semiconductors. Given the fact that the diffusion length is a critical parameter defining performance of p-n junction devices, it is imperative to find ways for its improvement. Our recent findings indicate that the latter parameter in ZnO can be noticeably enhanced by electron injection. The observed novel effect was attributed to electron trapping on impurity-related levels (10, 11).

Electron injection in bulk ZnO substrates

The experiments were carried out on commercially available (Tokyo Denpa (TD)) bulk ZnO substrates grown by hydrothermal technique. Secondary Ion Mass Spectroscopy (SIMS) measurements performed on these substrates revealed the presence of lithium (Li) in the crystal on the level of $\sim 4\times10^{16}$ cm^{-3} (Li is often added to ZnO to increase the resistivity of initially n-type samples) (12). Room temperature Hall measurements showed the samples to be a weak n-type with an electron concentration of $\sim 10^{14}$ cm^{-3} and mobility of ~ 150 cm^2/Vs. The samples under investigation were cleaved perpendicular to c-plane thus exposing the non-polar a-plane of ZnO. This was motivated by the observations that the latter crystallographic plane results in a better quality of Schottky contacts, as opposed to those deposited on the c-plane. Schottky barriers were, therefore, fabricated by electron beam evaporation of 100 nm-thick Au layer on ZnO a-plane and subsequent lift-off.

The experiments were carried out *in-situ* in a Philips XL30 Scanning Electron Microscope, which is integrated with a Gatan MonoCL3 cathodoluminescence system allowing wavelength-dependent and temperature-dependent optical measurements. A sample temperature in the cathodoluminescence measurements varied from 25 to 125 °C. For each temperature, periodic CL measurements were carried out under an SEM magnification of 4,000 at different locations subject to continuous (up to ~ 2200 seconds) excitation by an electron beam with the energy of 12 keV (fluence rate of $\sim 6\times10^{15}$ $cm^{-2}s^{-1}$), resulting in a current of up to several nA absorbed in the sample and electron beam penetration depth of ~ 0.8 μm.

CL results, obtained by periodic acquisition of cathodoluminescence spectra, were compared with those collected from the room temperature Electron Beam Induced Current measurements carried out on the same sample by moving an SEM electron beam from the edge of the Schottky barrier, outwards (line-scan), and recording an exponential decay of induced current. Detailed description of EBIC experiments can be found elsewhere (13). After a single EBIC line-scan was completed (12 seconds), the excitation of the sample was continued by moving the electron beam back and forth along the same line for the total time of ~ 2800 seconds. EBIC measurements were periodically repeated to extract the values of minority carrier diffusion length, **L**, as a function of the duration of electron beam irradiation, **t** (11, 13).

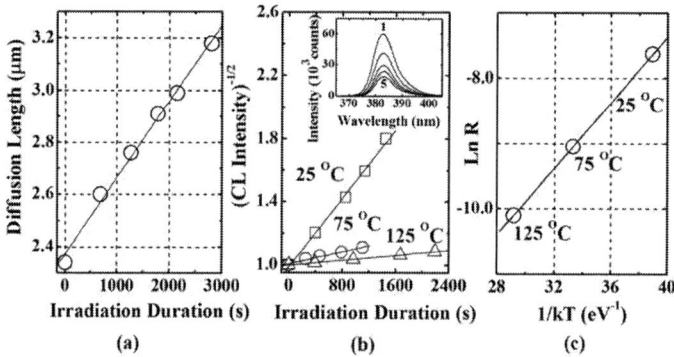

Fig. 1. a) Room temperature dependence of minority carrier diffusion length on duration of electron irradiation (open circles) and the linear fit (solid line). **b)** Variable-temperature dependence of inverse square root of normalized intensity on duration of electron irradiation and the linear fit with the rate R. **Inset:** Room temperature cathodoluminescence spectra taken under continuous excitation by the electron beam. **1** is the pre-irradiation spectrum and **5** is the spectrum after 1450 s of electron irradiation. **(c)** Arrhenius plot of R as a function of temperature yielding an activation energy $\Delta E_{A,I}$ of 283 ± 9 meV.

As displayed in Fig. 1a, a linear increase of **L** as a function of **t** is observed. **L** tends to saturate at **t** > 2800 seconds (not shown in Fig. 1a). We also note that the irradiation-induced increase of carrier diffusion length persists for at least 4 days at the same saturated level. The relatively large values of **L** in Fig. 1a (as compared to the values of ≤ 1 μm, usually observed in the direct band gap semiconductors) indicate very high quality and low dislocation density in bulk ZnO substrates.

EBIC measurements carried out on bulk ZnO substrates containing no lithium did not reveal any noticeable increase of **L** with **t** (up to 3600 s), suggesting that the presence of Li is important for the observed behavior (11).

The observed increase of **L** in bulk n-ZnO is attributed to an increase of lifetime, **τ**, (calculated variation of lifetime, based on measured values of **L**, are from ~ 0.4 to ~ 0.8 μs) for non-equilibrium minority holes in the valence band (due to a lower rate of recombination with non-equilibrium electrons). The experimental evidence for lifetime increase was obtained from CL measurements; the room temperature near band-edge luminescence in the inset of Fig. 1b exhibits a continuous decay with increasing duration of electron beam irradiation (injection).

As **L** is proportional to the square root of lifetime and depends linearly on **t** (14, 15, 16), the inverse CL intensity, **1/I**, which is also proportional to τ (larger τ ensures longer non-equilibrium minority carrier stay in the band, and, as a result, lower rate of radiative recombination), should depend on the duration of electron injection quadratically. This is, indeed, observed in Fig. 1b, where the square root of the inverse normalized (with respect to the initial maximum value) intensity, $1/\sqrt{I}$, is plotted versus **t**.

The rate, **R**, for the linear increase of $1/\sqrt{I}$ with duration of electron irradiation can be used to determine the activation energy of the irradiation-induced effect according to the following expression (11):

$$R = R_0 \exp\left(\frac{\Delta E_{A,I}}{kT}\right) \exp\left(-\frac{\Delta E_{A,T}}{2kT}\right) \qquad [1]$$

where R_0 is a scaling constant, **T** is temperature, **k** is the Boltzmann's constant, $\Delta E_{A,I}$ is the activation energy of electron irradiation effect, and $\Delta E_{A,T}$ is the activation energy of thermally-induced intensity decay determined as described in ref. (11). From the equation [1], the value for $\Delta E_{A,I}$ of 283 ± 9 meV was obtained using the Arrhenius plot shown in Fig. 1(c).

It is likely that the observed irradiation-induced increase in lifetime (cf. Fig. 1b) and diffusion length (cf. Fig. 1a) are caused by the trapping of non-equilibrium electrons on deep levels with the ionization energy of around 280 meV. Evidently, an increase in temperature results in a decrease of **R** (cf. Fig. 1c), suggesting existence of a thermally activated process, which counteracts the effects of electron irradiation, thereby leading to a slower rate of τ increase. Although more experiments are needed to clarify the nature of the involved deep center, it is quite likely related to the presence of Li in the ZnO lattice, as no electron injection effects occur in the material that does not contain lithium. It is pointed out in ref. (7) that group I elements substituting on the Zn site are good acceptor candidates. Li_{Zn}, for example, is predicted to have acceptor energy level of 90 meV. However, as is seen in TD substrates under investigation, Li doping actually produces highly resistive ZnO. The reason likely involves formation of Li-interstitials (which have a donor nature) and other point defects with a pronounced signature in the effect of electron injection. Several independent reports (17, 18) link the presence of Li to the formation of deep acceptor states in ZnO with thermal ionization energies consistent with that obtained by us for the effects of electron injection (~ 280 meV).

Electron injection in ZnO doped with antimony (ZnO:Sb)

The effects of electron injection for Sb-doped 0.2 μm-thick p-type ZnO epitaxial layers ($\rho = 1.3 \times 10^{17}$ cm^{-3}; $\mu = 8$ cm^2/Vs at room temperature) grown on Si substrate by Molecular Beam Epitaxy (MBE) are summarized in Fig. 2. The activation energy for the e-beam injection-induced increase of **L**, $\Delta E_{A,I} = 219 \pm 8$ meV, was obtained from the graphs in Fig. 2b,c using equation [1] (remembering that $L \sim 1/\sqrt{I}$) and accounting for the temperature-induced increase of diffusion length presented in Fig. 2a ($\Delta E_{A,T} = 184 \pm 10$ meV). The value of $\Delta E_{A,I}$ is in agreement with that for a $Sb_{Zn-2}V_{Zn}$ acceptor complex, predicted by Limpijumnong *et al.* (19).

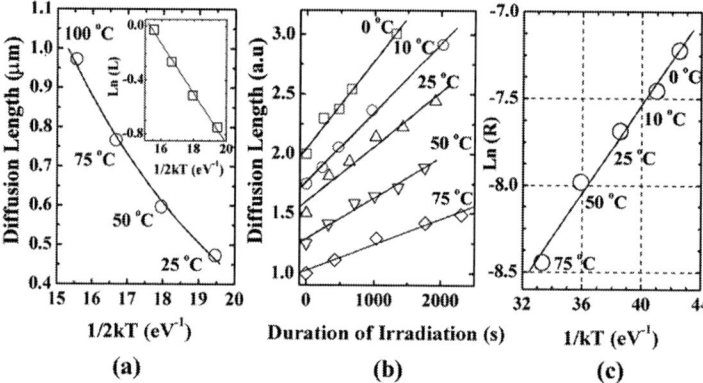

Fig. 2. a) Diffusion length of minority electrons as a function of temperature (open circles) and the fit (solid line). **Inset:** Arrhenius plot of the same data yielding activation energy of 184 ± 10 meV. **b)** Electron beam irradiation-induced increase of minority electron diffusion length at different temperatures. The values of the diffusion length were vertically offset for clarity and are not intended to illustrate the temperature dependence. **c)** Rate of irradiation-induced increase of diffusion length as a function of temperature (open circles). The fit with equation [1] (solid line) gives activation energy of 219 ± 8 meV. **See ref. (20) for details.**

It can also be seen from Fig. 2 that the rate, **R**, of the diffusion length increase is reduced with increasing temperature. The increase of the diffusion length due to trapping is counteracted by the release of the trapped electrons that occurs if the carriers gain sufficient energy to escape the trap. As the temperature is raised, the likelihood of de-trapping increases, which dampens the irradiation-induced growth of the diffusion length.

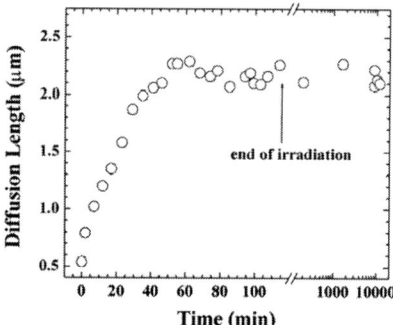

Fig. 3. Saturation and relaxation dynamics of minority carrier diffusion length in p-ZnO:Sb at room temperature. The arrow marks the time at which the electron irradiation was discontinued.

The saturation and relaxation of irradiation-induced change of diffusion length was studied at room temperature. Fig. 3 demonstrates that **L** reaches its maximum value after about 50 min of continuous exposure to the electron beam. Further monitoring revealed that irradiation-induced increase persists for at least one week. Annealing the sample at 175 °C for about 30 minutes resulted in a decrease of the diffusion length to about 1 μm.

Electron injection in ZnO p-n junction diodes under forward bias

If a p-n junction diode is biased in the forward direction, the junction potential barrier decreases and the electrons from the n-type region are injected into the p-type region. This is expected to result in an increase of minority carrier diffusion length in p-type ZnO and, therefore, enhancement of p-n junction's photoresponse.

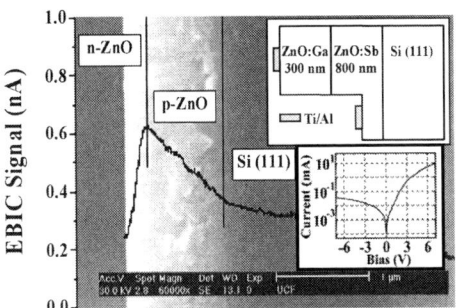

Fig. 4. Cross sectional image of ZnO p-n junction diode in secondary electrons and superimposed EBIC line-scan. **Inset (top):** schematic of ZnO epitaxial structure (not to scale). **Inset (bottom):** I-V curve of ZnO p-n junction.

Fig. 4 shows a secondary electron image and a superimposed EBIC line-scan of ZnO p-n epitaxial structure grown on highly resistive (20 $\Omega \cdot$cm) (111)-oriented p-Si substrate and cleaved perpendicular to the growth plane. The top inset of Fig. 4 presents a schematic of this structure. EBIC measurements on ZnO p-n junction (as well as spectral and temporal photoresponse measurements) were carried out at room temperature on the cleaved structures before and after forward bias electron injection (see I-V curve in the bottom inset of Fig. 4).

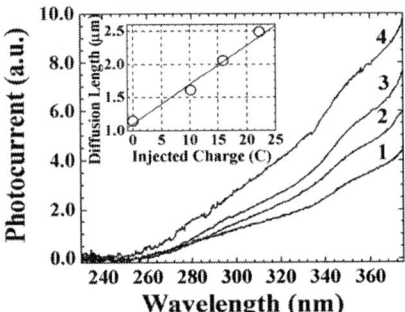

Fig. 5. Spectral response of ZnO p-n junction photodiode as a function of forward bias electron injection. Edge-illuminated configuration (as in Fig. 4) was employed for measurements. A shoulder observed on the spectra at ~ 350 nm corresponds to the effective band gap of ZnO (~ 3.54 eV). An increase of the photoresponse beyond this wavelength into visible region (ideally the response from device should end there) is due to the collection of photogenerated carriers in the Si substrate. Spectrum 1 corresponds to the pre-injected state; spectrum 2 – 2 C injected; spectrum 3 – 12.6 C injected; spectrum 4 – 25.5 C injected.
Inset: L dependence on charge in p-ZnO:Sb layer due to forward current injection. Note: this dependence was measured on p-n junction structure different from that showing a photoresponse in Fig. 5, but located on the same wafer as the latter. See ref. (21) for details.

The values of **L** in the p-type ZnO layer were extracted from the line-scans. The inset of Fig. 5 shows diffusion length dependence on injected charge (forward bias, resulting in currents from 7 to 48 mA, was applied for the duration of about 1500 seconds, in 300-600 seconds increment; **L** saturation was observed with increasing duration of injection, but it is not shown on the plot).

The exposure time of the p-n junction diode to the electron beam during EBIC measurements (12 seconds) is negligible as compared to that of forward bias application (up to ~ 1500 seconds). Therefore, the impact of electron beam injection on the increase of minority carrier diffusion length observed in the inset of Fig. 5 for p-ZnO layer is minimal. We also note that the diffusion length of minority holes in n-type ZnO (which was determined to be on the order of several hundred nm) was not affected by forward bias.

A significant increase of **L** with forward bias electron injection is consistent with a pronounced and long-lasting (persisted at the same level for at least several days!) enhancement of the spectral photoresponse of the ZnO p-n junction diode, as shown in Fig. 5. For lateral collection devices (side-illuminated, as in Fig. 4), the photocurrent is known to vary linearly with **L** (22).

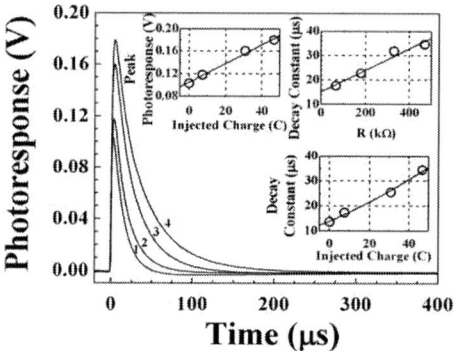

Fig. 6. Temporal response of ZnO p-n junction photodiode (cf. Fig. 4) as a function of forward bias electron injection. Upper left inset: dependence of peak photoresponse on injected charge. Upper right inset: decay constant of the improved photodiode (trace 4) as a function of external resistance (used for the measurements), indicating that the intrinsic decay time due to diode's series resistance is about 15 μs. Lower inset: dependence of decay constant on injected charge.

While an increase in **L** (due to longer lifetime) leads to a significant enhancement of quantum efficiency, the temporal photoresponse becomes slower as is seen in Fig. 6 for ZnO p-n junction diode excited by fs-pulse laser at 355 nm. Elongation of decay constant (Fig. 6, lower inset) is likely related to an increase of minority carrier lifetime due to electron injection. The result in Fig. 6 supports the general rule that more sensitive detectors are slower (23).

Discussion

If photons are absorbed on the p-side of a photodetector, the quantum efficiency, η, is (24):

$$\eta = (1-R)\left(1 - \frac{e^{-\alpha W}}{1+\alpha L}\right) \qquad [2]$$

here, $R(\alpha)$ is the reflection (absorption) coefficient, W is the intrinsic (i) (depletion) layer width, and L is the p-side minority electron diffusion length. Increasing L clearly improves η and responsivity (proportional to quantum efficiency (24)).

The tentative model for the observed electron injection-induced effects is presented in Fig. 7 (13, 25, 26).

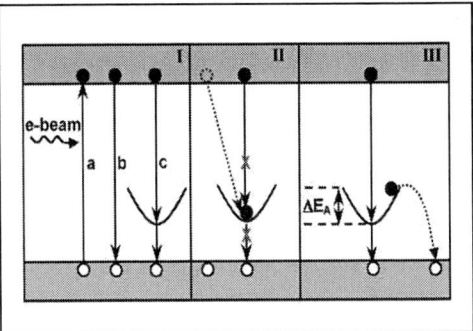

Fig. 7. Model of the electron injection effect induced by an electron beam irradiation. Electron beam, e-beam, generates non-equilibrium electron-hole pairs (**Ia**). Non-equilibrium carriers recombine either via the band-to-band transition (**Ib**) or through unoccupied (non-ionized) acceptor states (**Ic**). However, if a non-equilibrium electron is trapped by the acceptor level, recombination cannot proceed (**II**), leading to increased lifetime of non-equilibrium carriers. Release of the trapped electron with an activation energy ΔE_A (**III**) restores the original recombination pathway (**Ic**), resulting in a slower rate of lifetime increase at elevated temperatures.

The key ideas of the model are summarized below:
- A non-equilibrium electron, generated by a Scanning Electron Microscope (SEM) beam (cf. Fig. 7,I) (or due to a forward bias of p-n junction), is trapped by a neutral meta-stable level (cf. Fig. 7,II). The concentration of involved dopant levels ($\sim 10^{18}$ cm^{-3}) increases with the duration of electron injection. Trapping of the non-equilibrium electrons on the impurity levels (these levels create a band in the semiconductor forbidden gap) prevents recombination of the conduction band electrons through these levels (cf. Fig. 7,II). This leads to an increase of lifetime, τ, for a non-equilibrium electron in the conduction band and, as a result, an increase of L ($L = \sqrt{D\tau}$, where D is carrier diffusivity which is unaffected by electron beam irradiation (13)) and a decrease of CL intensity (radiative recombination rate is inversely proportional to carrier lifetime).
- The impurity level containing a trapped electron becomes available for recombination of a non-equilibrium conduction band electron as this level captures a hole. Capturing a hole means a transition of the trapped electron to the valence band (cf. Fig. 7,III). The rate of this transition increases with increasing temperature, and we note the existence of the activation energy, preventing the immediate hole capture by the ionized impurity. It has to be pointed out that the activation energy for a hole capture on the impurity level is comparable (within 20%) with the thermal ionization energy of this level (14).
- As the rate of hole capture on the impurity level increases, the conduction band electrons have more chance for recombination on this level. This results in a slower rate for the increase of non-equilibrium minority electron lifetime and L at higher temperatures.

Conclusions

Electron injection into ZnO semiconductors leads to a pronounced enhancement of minority carrier transport (diffusion length) due to longer minority carrier lifetime in the band. These effects are attributed to electron trapping on deep levels in the ZnO forbidden gap.

Acknowledgments

This research is supported in part by the National Science Foundation (ECCS #0900971) and US-Israel Binational Science Foundation (award # 2008328).

References

1. O. Lopatiuk, W.C. Burdett, L. Chernyak, K.P. Ip, Y.W. Heo, D.P. Norton, S.J. Pearton, B. Hertog, P.P. Chow, A. Osinsky, *Appl. Phys. Lett.*, **86**, 012105 (2005).
2. O. Lopatiuk, L. Chernyak, A. Osinsky, J.Q. Xie, *Appl. Phys. Lett.*, **87**, 214110 (2005).
3. O. Lopatiuk-Tirpak, L. Chernyak, F.X. Xiu, J.L. Liu, S. Jang, F. Ren, S.J. Pearton, K. Gartsman, Y. Feldman, A. Osinsky, P. Chow, *J. Appl. Phys.*, **100**, 086101 (2006).
4. O. Lopatiuk, A. Osinsky, L. Chernyak, , in Zinc Oxide Bulk, Thin Films and Nanostructures, C. Jagadish and S. Pearton, Editors, Elsevier Ltd. (2006).
5. O. Lopatiuk-Tirpak, L. Chernyak, L.J. Mandalapu, Z. Yang, J.L. Liu, K. Gartsman, Y. Feldman, Z. Dashevsky, *Appl. Phys. Lett.*, **89**, 142114 (2006).
6. O. Lopatiuk-Tirpak, G. Nootz, E. Flitsyian, L. Chernyak, L.J. Mandalapu, Z. Yang, J.L. Liu, K. Gartsman, A. Osinsky, *Appl. Phys. Lett.*, **91**, 042115 (2007).
7. D.C. Look, B. Claflin, Ya.I. Alivov, S.J. Park, *Phys. Stat. Sol. (A)*, **201**, 2203, (2004).
8. Z.P. Wei , Y.M. Lu , D.Z. Shen , Z.Z. Zhang , B. Yao, B.H. Li, J.Y. Zhang, D.X. Zhao, X.W. Fan, Z.K. Tang, *Appl. Phys. Lett.*, **90**, 042113 (2007).
9. L.J. Mandalapu, Z. Yang, F.X. Xiu, D.T. Zhao, J.L. Liu, *Appl. Phys. Lett.*, **88**, 092103 (2006).
10. O. Lopatiuk, W.C. Burdett, L. Chernyak, K.P. Ip, Y.W. Heo, D.P. Norton, S.J. Pearton, B. Hertog, P.P. Chow, A. Osinsky, *Appl. Phys. Lett.*, **86**, 012105 (2005).
11. O. Lopatiuk, L. Chernyak, A. Osinsky, J.Q. Xie, *Appl. Phys. Lett.*, **87**, 214110 (2005).
12. A.Y. Polyakov, N.B. Smirnov, A.V. Govorkov, E.A. Kozhukhova, S.J. Pearton, D.P. Norton, A. Osinsky, A. Dabiran, "Electrical properties of undoped bulk ZnO substrates", *J. Electron. Materials*, **35**, 663-669 (2005).
13. L. Chernyak, A. Osinsky, V. Fuflyigin, E.F. Schubert, "Electron beam-induced increase of electron diffusion length in p-type GaN and AlGaN/GaN superlattices", *Appl. Phys. Lett.*, **77**, 875-877 (2000).
14. L. Chernyak, W. Burdett, *MRS Spring Meeting Proceedings*, **764**, C5.4 (2003).
15. L. Chernyak, W. Burdett, M. Klimov, A. Osinsky, *Appl. Phys. Lett.*, **82**, 3680, (2003).
16. W. Burdett, O. Lopatiuk, A. Osinsky, S.J. Pearton, L. Chernyak, *Superlattices and Microstructures*, **34**, 55, (2004).
17. B.K. Meyer, H. Alves, D.M. Hofmann, W. Kriegseis, D. Forster, F. Bertram, J. Christen, A. Hoffmann, M. Straßburg, M. Dworzak, U. Haboeck, A.V. Rodina, *Phys. Stat. Sol. B*, **241**, 231, (2004).
18. M.G. Wardle, J.P. Goss, P.R. Briddon, *Phys. Rev. B*, **71**, 155205 (2005).

19. S. Limpijumnong, S. B. Zhang, S. H. Wei, C. H. Park, *Phys. Rev. Lett.*, **92**, 155504 (2004).
20. O. Lopatiuk-Tirpak, L. Chernyak, F.X. Xiu, J.L. Liu, S. Jang, F. Ren, S.J. Pearton, K. Gartsman, Y. Feldman, A. Osinsky, P. Chow, *J. Appl. Phys.*, **100**, 086101 (2006).
21. O. Lopatiuk-Tirpak, L. Chernyak, L.J. Mandalapu, Z. Yang, J.L. Liu, K. Gartsman, Y. Feldman, Z. Dashevsky, "Influence of electron injection on the photoresponse of ZnO homojunction diodes", *Appl. Phys. Lett.*, **89**, 142114 (2006).
22. H. Holloway, "Theory of lateral-collection photodiodes", *J. Appl. Phys.*, **49**, 4264-4269 (1978).
23. L. Chernyak, A. Schulte, A. Osinsky, *Encyclopedia of Sensors*, C.A. Grimes and E.C. Dickey, Editors, *American Scientific Publishers*, **Vol. X**, 1-14 (2006).
24. S.M. Sze, *Semiconductor Devices Physics and Technology*, Wiley, New York (1985).
25. L. Chernyak, A. Osinsky, A. Schulte, *Solid State Electron.*, **45**, 1687 (2001).
26. W. Burdett, A. Osinsky, V. Kotlyarov, P. Chow, A. Dabiran, L. Chernyak, *Solid-State Electron.*, **47**, 931 (2003).

12

Aluminum-Doped Zinc Oxide Thin Films for Opto-Electronic Applications

N. Hirahara[a], B. Onwona-Agyeman[a] and M. Nakao[a]

[a] Kyushu Institute of Technology, Kitakyushu, Fukuoka 804-8550, Japan

Transparent conducting aluminum-doped Zinc oxide (AZO) thin films were prepared on glass substrates by rf magnetron sputtering technique using ZnO ceramic target in pure argon gas with different aluminum concentrations. The bandgap of the ZnO films slightly widens with increase in Al content and the lowest sheet resistance of AZO films with Al concentration of 4.25 atomic % was obtained. The effects of post-annealing treatment on structural, electrical and optical properties of the AZO thin films were investigated. Using AZO film with 4.2 at. % Al as the transparent electrode, a titanium dioxide based dye-sensitized solar cell was constructed and a solar to electrical energy conversion efficiency of 2.9 % was achieved under AM 1.5 solar simulated sunlight.

Introduction

Transparent conductive oxide (TCO) films such as tin-doped indium oxide (ITO), fluorine-doped tin oxide (FTO) and aluminum-doped zinc oxide (AZO) have high transmittance in the visible region of the electromagnetic spectrum combined with reasonable conductivity. TCO films have therefore been used extensively as components in opto-electronic devices with applications in solar cells, light emitting diodes, flat and touch panel displays (1-3). ITO and FTO contain indium and tin which are rare earth metals making ITO and FTO expensive TCO. Zinc oxide on the other hand has been actively investigated as an alternative because it is cheaper, non-toxic and readily available. Impurity-doped ZnO is more appropriate for TCO applications because the electrical resistivity of undoped ZnO is not low enough to be used as TCO (4, 5). Doping ZnO with Group III elements (Al, In and Ga) drastically lowers the electrical resistivity (6, 7) of the ZnO material because the dopant element partially replaces Zn thereby creating additional energy levels. These energy levels play significant roles in the observed optical transmittance and therefore it is important to optimize the concentration of the dopant element in the ZnO film.

Many opto-electronic devices require thermal treatments during fabrication and therefore transparent conductive oxides (TCOs) must maintain its properties during such high temperature process. Dye-sensitized solar cell (DSC) is an opto-electronic device that uses TCO as transparent electrode and requires thermal treatment during fabrication (8, 9). In the DSC, usually the nano-porous oxide semiconductor is deposited on the TCO and heated at 500 °C to remove organic contaminants and to create enough pores for efficient dye adsorption. However, the resistivity of glass-ITO can increase significantly after thermal treatment therefore affecting the performance of the DSC. Aluminum-doped ZnO can withstand this thermal treatment without significant change in transmittance and resistivity.

In this work, we have prepared AZO thin films using rf magnetron sputtering technique. The aluminum concentration was systematically increased and the resistivity and optical transmission were investigated. Finally using the AZO thin films as transparent electrode, a DSC was constructed and it's light to electrical energy conversion properties evaluated.

Experimental

The aluminum-doped ZnO thin films were deposited on quartz-glass substrates by radio-frequency magnetron sputtering. The substrates were ultrasonically cleaned in acetone, water and ethanol, dried before loading into the growth chamber. A commercially available pure ZnO ceramic target 50 mm in diameter (purity 99.99%) was used as the target and the distance between the target and substrate was kept at 50 mm. The growth chamber was evacuated to 4×10^{-4} Pa by turbo-molecular and rotary pumps before introducing argon as the sputtering gas. During deposition, the sputtering pressure was maintained at 1.5 Pa, rf power at 120 W and the substrate temperature at 300 °C.

To introduce aluminum doping, aluminum foils (purity ~99.999%, 0.1 mm in thickness, Nilaco Corporation, Japan) were placed on the effective sputtering area of the ZnO target during deposition and the area of the Al foil was varied to control the Al content in the ZnO films. The concentration of the Al in the ZnO film was determined by Energy Dispersive X-ray Spectrophotometer (EDX). The Al doping concentration was 3.78, 4.25, 5.05 and 5.72 at. % respectively and the average film thickness was 2.3 μm. The crystal structure of the ZnO films was evaluated by an X-ray diffractometer (JEOL JDX-3500K) using Cu Kα and the optical transmission spectra were measured using UV/Vis spectrometer (JASCO V560) and all transmission spectra were taken at room temperature in air. The film thickness was determined by Scanning Electron Microscopy (FE-SEM JSM-7000 FSK) and the film surface roughness was observed by Atomic Force Microscopy (Toyo Technica). The DSC was constructed by depositing TiO$_2$ paste (Ti-Nanoxide HT/SP) on the AZO-glass substrate by screen printing and heating at 500°C for 30 min. in air. The TiO$_2$ films were then coated with a ruthenium dye (N-719) by dipping the films in dye solutions for about 12 hours. The dye-coated TiO$_2$ electrode was sandwiched with a Pt-sputtered conducting glass plate as counter electrode and the intervening space filled with an electrolyte. I-V characteristics (1.5 AM 100 Wm^{-2} simulated sunlight) were recorded with a calibrated solar-cell evaluation system (Yamashita Denso YSS-50) and the cell active area was 0.25 cm^2.

Results and Discussion

Fig. 1 shows the XRD patterns of the undoped and Al-doped ZnO films deposited on quartz substrates. The (002) diffraction peak appeared in all the films indicating differential c-axis orientation of the ZnO films. In addition to the (002) peak, the undoped ZnO films shows different diffraction peaks but these peaks disappeared with increasing Al concentration. The AZO film with 3.78 at. % of Al show the (100), (101) and (103) peaks in addition to the (002) peak. With Al concentration of 4.25, 5.05 and 5.72 at. %, only the (002) peak appeared in the XRD patterns. The average grain sizes of the films were calculated from the value of the Full width at Half Maximum (FWHM) of the (002) diffraction peak using the Scherrer's relation (10). From the XRD analysis, the ZnO films with 4.25 atomic % concentrations has the largest grain size with the narrowest FWHM of the (002) indicating good c-axis orientation.

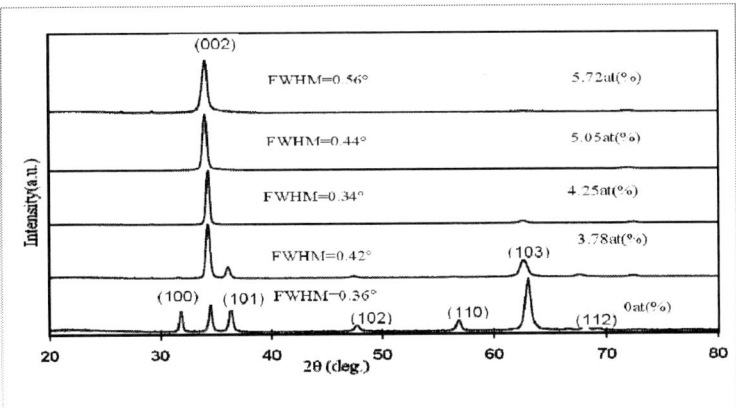

Figure 1. XRD patterns of ZnO and AZO films with different Al concentrations annealed in air at 500 °C

Table 1 summarizes the sheet resistance and optical transmission data for the undoped ZnO and AZO films after heating in air at 500 °C and Fig. 2 shows the optical transmittance spectra of the undoped and AZO films heated at 500 °C in the wavelength range between 350 and 850 nm. To be used effectively as a transparent electrode in a DSC, the AZO films must retain its properties (sheet resistance and transmittance) after heating at about 500 °C. It can be observed from Table 1 that, the sheet resistance of the AZO films did not change much after the thermal treatment and the lowest sheet

TABLE I. Sheet resistance and optical transmittance of as-grown and annealed (500 °C) ZnO and AZO films

Al content (at. %)	Sheet Resistance (Ω/sq)		Optical Transmittance (%)	
	before	after	before	after
0	X	X	80	81
3.78	10.1	8.2	74	76
4.25	4.9	4.8	78	79
5.02	10.9	14.2	83	85
5.77	17.5	18.7	84	85

resistance of 4.8 Ω/sq and average transparency of 79 % was obtained with Al concentration of 4.25 at. %. The sheet resistance of the undoped film did not improve even after heating but the transparency was about 81 % in the visible range. The optical transmittance spectra shows that all the AZO films including the undoped ZnO increased after heating at 500 °C. To be used as a transparent electrode, the AZO films must have low sheet resistance and significant transparency. From the results in Table 1, the AZO film with 4.25 at. % Al can be considered the most suitable to be used as the transparent electrode in the DSC.

In order to calculate the bandgap of the undoped and doped ZnO films, we used the Tauc`s relationship (11) as follows;

$$\alpha h\nu = A \, (h\nu - E_g)^n, \qquad [1]$$

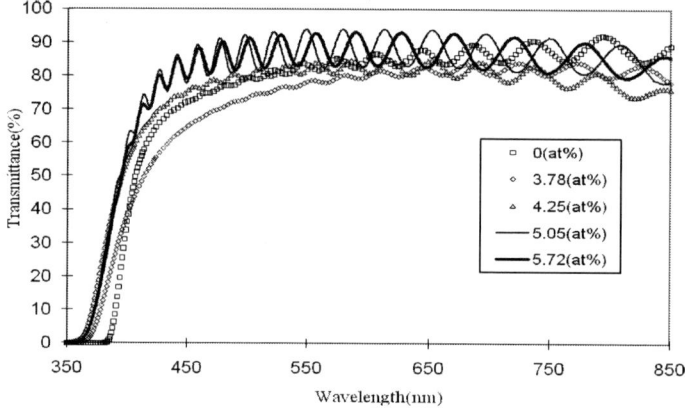

Figure 2. Transmittance spectra for Al doped ZnO with different Al concentrations annealed in air at 500 °C

where α is the absorption coefficient, A, a constant, h is Planck's constant, v the photon frequency, E_g the optical bandgap and n is ½ for a direct bandgap semiconductor. An extrapolation of the linear region of a curve of $(\alpha h v)^2$ on the y-axis against the photon energy (hv) on the x-axis gives the value of the bandgap E_g. According to the absorption spectra measured, $(\alpha h v)^2$ versus hv curves of undoped and doped ZnO films were plotted and the bandgap values were evaluated. Figure 3 shows the dependence of the bandgap values on the doping concentration of Al in the ZnO films. Bandgap values of 3.217, 3.402, 3.403 and 3.407 eV were obtained for the undoped and doped ZnO films. It can be seen that the energy bandgap of the undoped ZnO film is about 3.217 eV and this increases with the increase in the Al content. This means that the bandgap widens with the introduction of Al.

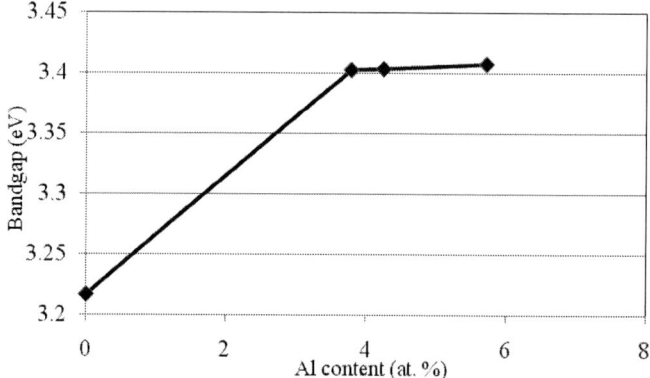

Figure 3. Dependence of bandgap values on Al doping of ZnO films

Since the AZO film with Al content of 4.25 at. % gave the lowest sheet resistance, a TiO_2-based DSC was prepared using this AZO film as the transparent electrode instead of the usual FTO and ITO. Fig. 4 shows SEM and AFM images of the surface morphology of AZO films with 4.25 at. % Al content. The AFM image was observed using a scale of 10 µm x 10 µm and the root-mean-square (RMS) surface roughness value of about 23 nm was obtained. The RMS of the undoped film was 27 nm which is higher than the AZO film with 4.25 at. % Al. From the XRD patterns in Fig. 1, the undoped ZnO film shows many orientations, which means the film is composed of many grains with different crystallographic orientations whilst the AZO film with 4.25 at. % Al has mainly the (002) orientation. In general, the surface of polycrystalline films is rougher due to the random orientation of the grains (12) and therefore the AZO films with (002) orientation has smaller RMS value than the undoped film. The high RMS values of the undoped and AZO films may be due to the fact that, the substrates were not rotated during sputtering.

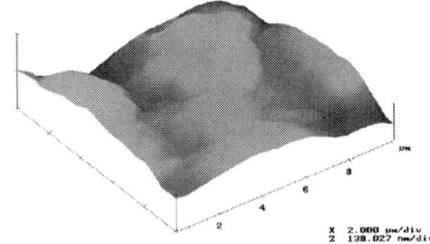

Figure 4. SEM and AFM images of ZnO film doped with 4.25 at % Al

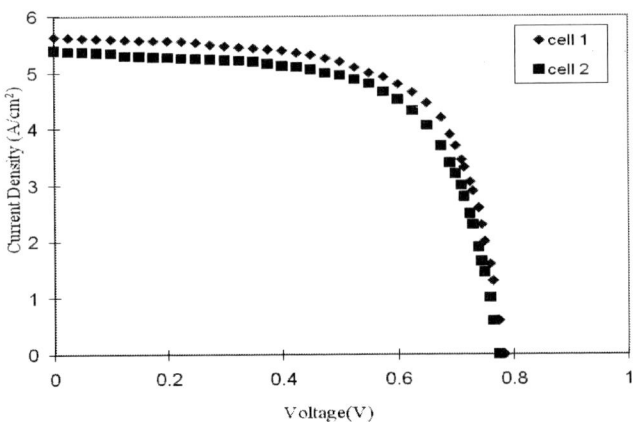

Figure 5. *I-V* characteristics of two TiO_2-based DSC with AZO film as the transparent electrode

TABLE II. I-V parameters (J_{sc} = short-circuit photocurrent, V_{oc} = open-circuit voltage, FF = fill factor) of TiO_2 DSC with AZO films as transparent electrode

Solar cell	Voc (volts)	Jsc /mAcm$^-$^{2}	FF	Efficiency/%
1	0.785	5.64	0.656	2.91
2	0.775	5.40	0.648	2.70

The DSC was prepared by depositing the TiO_2 paste on two separate AZO films (4.25 at. % Al) for comparison and heating in air at 500 °C. The calcined TiO_2 films were then sensitized by immersing in a solution containing the ruthenium dye (N719) for 12 h. The two DSC were assembled using the TiO_2 on the AZO as the photoanode, Pt sputtered on glass as photocathode and introducing an electrolyte to complete the DSC structure (13). The photocurrent-voltage characteristics were measured using a solar simulator. Figure 5 shows the *I-V* characteristics of the two DSC using the AZO films as the transparent electrode and Table 2 summarizes the *I-V* parameters of the two DSC. The conversion efficiencies of the two cells prepared under the same experimental conditions were 2.91 % and 2.70 % respectively both lower than when FTO is used as the transparent electrode (14, 15). The voltage and fill factors (FF) recorded in this work are equivalent to of FTO-based DSC systems. Reproducibility of the conversion efficiency of the two cells indicates that, AZO films can be successfully used as transparent electrode in DSC. By optimizing certain sputtering parameters such as sputter pressure and rf power, improving the surface roughness of the AZO films used in this work and also reducing the thickness of the AZO films may all have significant effect on the conversion efficiency of the DSC. Also, the low photocurrents observed may be due to poor adsorption of dye by the TiO_2 particles which affected the conversion efficiency.

Conclusions

We have successfully prepared aluminum-doped ZnO thin films on glass substrates using rf magnetron sputtering. The effects of Al doping concentration on the sheet resistance and transmittance of the ZnO films were investigated. Using AZO film with 4.25 at. % Al as the transparent electrode, a TiO_2-based dye-sensitized solar cell was successfully prepared. Efficiency of 2.91 % recorded for the AZO film as transparent electrode in TiO_2-based DSC with an area of 0.25 cm^2 under AM 1.5 irradiation is significant and proves the potential of AZO films in dye-sensitized solar energy conversion.

Acknowledgments

The authors are grateful to Prof. R. Shiratsuchi of Kyushu Institute of Technology for technical support for the dye-sensitized solar cells, Prof. Y. Sun of Kyushu Institute of Technology for the optical transmission spectra measurements and also to Prof. T. Haruyama of Research Center for Advanced Eco-fitting Center, Kyushu Institute of Technology.

References

1. C.G. Granqvist, Solar *Energy Mater. and Solar Cells,* **91**, 1529 (2007).
2. P.S. Patil, R.K. Kawar, S.B. Sadale, P.S. Chigare, *Thin Solid Films*, **437**, 34 (2003).
3. A. Dima, O. Dima, C. Moldovan, C. Cobianu, C. Savaniu and Zaharescu, *Thin Solid Films*, 427, 427 (2003).
4. S. Mayer and K.L. Chopra, *Sol. Energy Mat.,* **17**, 319 (1998).
5. H.A. Wanka, E. Lotter and M.B. Schubert, Mater. *Res. Soc. Symp. Proc.*, **336**, 657 (1994).
6. T. Minami, H. Nanto and S. Takata, *Jpn. J. Appl. Phys.*, **24**, L.781 (1985).
7. J. Ma, F. Ji, D.H. Zhang, H.L. Ma, S.Y. Li, *Thin Solid Films*, **347**, 1 (199).
8. K. Kalyanasundaram and M. Grätzel, *Coord. Chem. Rev.,* **77**, 347 (1998).
9. M. Grätzel, *Curr. Opin. Colloid. Interface Sci.,* **4**, 314 (1999).
10. L.V. Azaroff, *Elememt of X-ray Crystallography*, McGraw-Hill, New York (1968).
11. J. Tauc (Ed), *Amorphous and Liquid Semiconductors*, Plenum, New York (1974).
12. K. Yim and C. Lee, *Cryst. Res. Technol,*. **41**, No. 12, 1198 (2006).
13. J. Yamamoto, A. Tan, R. Shiratsuchi, S. Hayase, C. Ramannai and K. Rajeshwar, *Advanced Materials,* **15**, 1823, (2003).
14. M.K. Nazeeruddin, P. Péchy, T. Renouard, S.M. Zakeeruddin, R. Humphry-Baker, P. Comte, P. Liska, L. Cevey, V. Shklover, L. Spiccia, G.B. Deacon, C.A. Bignozzi and M. Grätzel, *J. Am. Chem. Soc.,* **123**, 1613 (2001).
15. Y. Chiba, A. Islam, Y. Watanabe, R. Komiya, N. Koide and L. Han, *Jap. J. Appl. Phys.*, **45**, L638 (2006).

20

ECS Transactions, 28 (4) 21-26 (2010)
10.1149/1.3377095 ©The Electrochemical Society

Investigation and Fabrication of Bottom Gate ZnO:Al TTFTs with Various Thicknesses of ZnO Buffer Layers

Yung-Hao Lin[a], Hsin-Ying Lee[b], and Ching-Ting Lee[a]

[a] Institute of Microelectronics, Department of Electrical Engineering,
National Cheng Kung University, Tainan, 701, Taiwan, Republic of China
[b] Department of Electro-Optical,
National Cheng Kung University, Tainan, 701, Taiwan, Republic of China

An Al doped ZnO (ZnO:Al) transparent thin film transistors
(TTFTs) with various thickness of ZnO buffer layer sandwiched
between gate insulator and channel layer were deposited by a
magnetron radio frequency co-sputter system. When the thickness
of the buffer layer was 80 nm, the field-effect carrier mobility of
the TTFTs was as high as 122.0 cm^2/V-s. Furthermore, the
associated gate voltage swing was 0.24 V/decade, and the
maximum state density was 2.69×10^{11} $eV^{-1}cm^{-2}$. The on-to-off
current ratio of the TTFTs with 80 nm-thick ZnO buffer layer was
up to 5×10^7.

Introduction

Recently, new technologies of flat panel display have been put forth continually
because of the advanced development of thin film transistor liquid crystal display (TFT
LCD). At the present, the TFTs available in the market are mostly made of amorphous Si
(a-Si) and low temperature poly-Si film. However, TFTs with a-Si channel layers will
produce photocurrent under illumination and hence affect the dark current of the
transistor. The photocurrent decreased the display quality. To overcome this problem, a
black matrix was used to isolate the TFTs from the panel back light. However, it would
reduce the aperture ratio. In previous research, ZnO and ITO were used to fabricate
TTFTs [1-4], because these materials do not absorb visible light via the wide band gap
materials. In this work, to avoid the problems of the lattice mismatch between channel
layer and gate insulator and to obtain a better crystallinity of the ZnO:Al channel layer,
various-thick ZnO buffer layers sandwiched between Al-doped ZnO (ZnO:Al) channel
layer and gate insulator were deposited. The electrical characteristics of the ZnO:Al
TTFTs fabricated with various thicknesses of ZnO buffer layers, were discussed.

Experimental Procedure

The cross-sectional configuration of the bottom gate ZnO:Al TTFTs with the ZnO
buffer layer is shown in Figure 1. The layers of TTFTs were deposited using a magnetron
RF co-sputtering system. A 200 nm-thick SiO_2 gate insulator was first deposited on ITO
glass substrates. Various thicknesses of 0, 50, 80, and 200 nm-thick ZnO buffer layer
were then deposited on the SiO_2 gate insulator. A 25 nm-thick ZnO:Al channel layer was
followed to deposit on the ZnO buffer layer. For making the high ohmic performance, a
20 nm-thick n^+-ZnO (Al doped) layer was deposited on the ZnO:Al channel layer. The
source and drain electrodes were fabricated by a 200 nm-thick ITO layer. The channel
width (W) and channel length (L) were 100 μm and 10 μm, respectively. The grain size of

21

the deposited ZnO:Al channel layer with ZnO buffer layer was estimated as 21.7 nm which is larger than 19.4 nm of that without ZnO buffer layer. The insertion of the ZnO buffer layer was intended to improve the quality of the ZnO:Al channel layer. The electrical characteristics of the ZnO:Al TTFTs with various thicknesses of ZnO buffer layer were measured using Agilent 4156C semiconductor parameter analyzer.

Experimental Results and Discussion

The drain-source current (I_{DS})-gate-source voltage (V_{GS}) characteristics of the ZnO:Al TTFTs with (a)0, (b)50, (c)80, and (d) 200 nm-thick ZnO buffer are shown in Figure 2, 3, 4, and 5, respectively. The field-effect carrier mobility (μ_{FE}) in the triode region can be estimated by using eq. (1) [5]:

$$I_{DS} = \frac{W\mu_{FE}C_{OX}}{2L}\left[2(V_{GS}\text{-}V_{T})V_{DS}\text{-}V_{DS}^2\right] \tag{1}$$

where C_{ox} is the capacitor of unit area gate insulator, and V_T is the threshold voltage. As inserting various thicknesses of ZnO buffer layers, the effective field-effect carrier mobility of the ZnO:Al TTFTs with 0, 50, 80, 200 nm-thick ZnO buffer layer calculated from Figure 2, 3, 4, and 5 is 32.5, 80.8, 122.0, and 29.1 cm²/V-s, respectively. The ZnO:Al TTFTs exhibit a better electronic performance as inserting 80 nm-thick ZnO buffer layer which resulted in the quality of the ZnO:Al channel layer was improved by inserting ZnO buffer layer. The increase in effective field-effect carrier mobility is attributed to the increase in the grain size of the channel layer with ZnO buffer layer from 19.4 nm to 21.7 nm. Nevertheless, the effect electric field in the ZnO:Al channel layer is reduced by a too thick ZnO buffer layer ,and makes a decrease of effective field-effect carrier mobility.

The on-to-off current ratio of the ZnO:Al TTFTs with 80 nm-thick ZnO buffer layer is up to 5×10^7 which is better than 2.8×10^7 of the one without ZnO buffer layer. The gate voltage swing (S) and the maximum state density (N_{ss}^{max}) of the ZnO:Al TTFTs were estimated by using the following eq. (2) and eq. (3), respectively [6, 7]:

$$S = \frac{dV_{GS}}{d\log I_{DS}} \tag{2}$$

$$N_{ss}^{max} = \left(\frac{S\log e}{kT/q}-1\right)\frac{C_{ox}}{q} \tag{3}$$

where k is the Boltzmann constant, q is the electron charge, T is the absolute temperature, and e is the base of the natural logarithm. The S value of the ZnO:Al TTFTs with 0, 50, 80,and 200 nm-thick ZnO buffer layer is 0.39, 0.30, 0.24, and 0.27 V/decade, respectively. Furthermore, the associated N_{ss}^{max} value is 5.72×10^{11}, 3.76×10^{11}, 2.69×10^{11}, and 2.58×10^{11} eV⁻¹cm⁻², respectively. The results shown in Figure 6 indicate the ZnO buffer layer improves the crystalline structure of the ZnO:Al TTFTs.

Conclusions

In this work, a ZnO:Al TTFTs with various thickness of ZnO buffer layer were deposited using a magnetron RF co-sputtering system. The grain size of the ZnO:Al channel layer increased by inserting the ZnO buffer layer. The inserted ZnO buffer layer could improve the crystal quality of the ZnO:Al channel layer. The effective field-effect carrier mobility of the ZnO:Al TTFTs with a ZnO buffer layer increased from 32.5 to 122.0 cm^2/V-s. It also led the gate voltage swing decrease from 0.39 to 0.24 V/decade and led the maximum state density decrease from 5.72×10^{11} to 2.69×10^{11} $eV^{-1}cm^{-2}$. In conclusions, ZnO-based TTFTs exhibit a better performance by sandwiching a proper thickness of ZnO buffer layer.

Acknowledgments

This work was supported by the National Science Council of Taiwan and Chi Mei Optoelectronics Corp., Taiwan, Republic of China.

References

1. H. Q. Chiang, J. F. Wager, R. L. Hoffman, J. Jeong, and D. A. Keszler, *Appl. Phys. Lett.* **86**, 013503 (2005).
2. J. M. Choi, D. K. Hwang, J. H. Kim, and S. Im, *Appl. Phys. Lett.* **86**, 123505 (2005).
3. N. L. Dehuff, E. S. Kettenring, D. Hong, H. Q. Chiang, J. F. Wager, R. L. Hoffman, C.-H. Park, and D. A. Keszler, *J. Appl. Phys.* **97**, 064505 (2005).
4. E. Fortunato, P. Barquinha, G. Gonçalves, L. Pereira, and R. Martins, *Solid-State Electron.*, **52**, 443-448 (2008).
5. M. Shur, M. Hack and J. G. Shaw, *J. Appl. Phys.* **66**, 3371 (1989).
6. R. Martins, P. Barquinha, I. Ferreira, L. Pereira, G. Gonçalves, and E. Fortunato, *J. Appl. Phys.* **101**, 044505 (2007).
7. A. Rolland, J. Richard, J. P. Kleider and D. Mencaraglia, *J. Electrochem, Soc.* **140**, 3679 (1993).

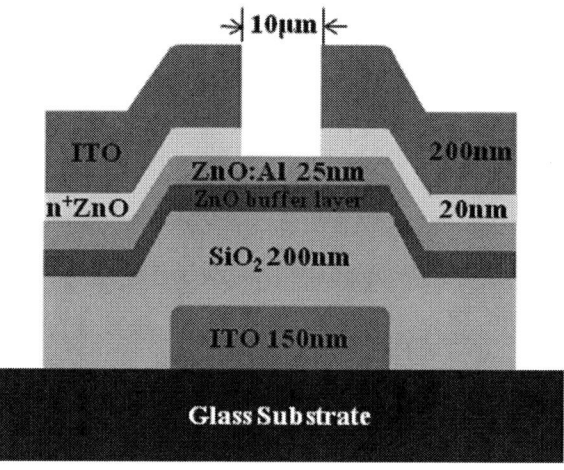

Figure 1. The schematic cross-sectional configuration of bottom gate ZnO:Al TTFTs with ZnO buffer layer.

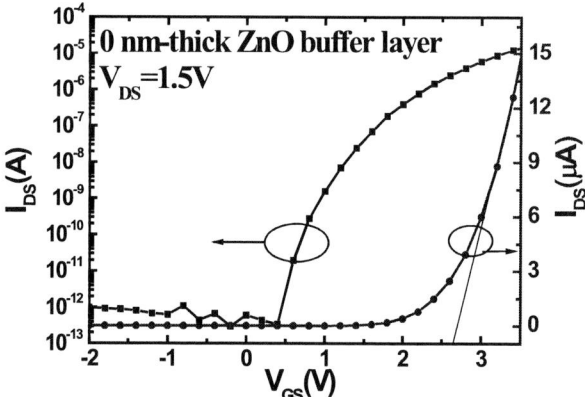

Figure 2. The I_{DS}-V_{GS} characteristics of ZnO:Al TTFTs without ZnO buffer layer.

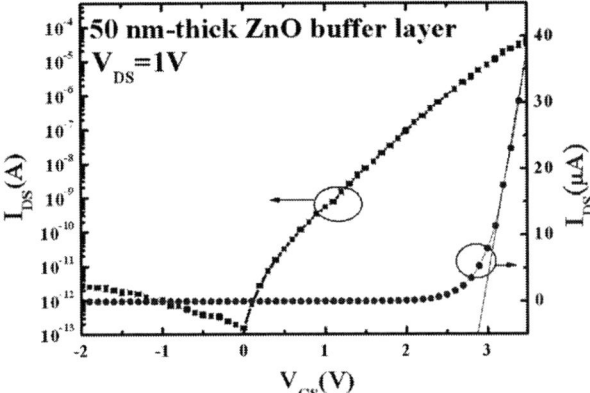

Figure 3. The I_{DS}-V_{GS} characteristics of ZnO:Al TTFTs with 50 nm-thick ZnO buffer layer.

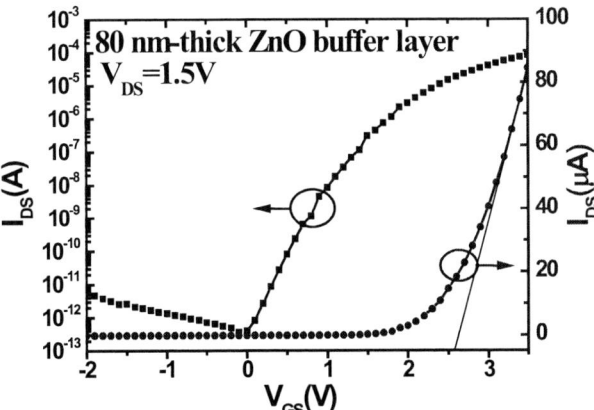

Figure 4. The I_{DS}-V_{GS} characteristics of ZnO:Al TTFTs with 80 nm-thick ZnO buffer layer.

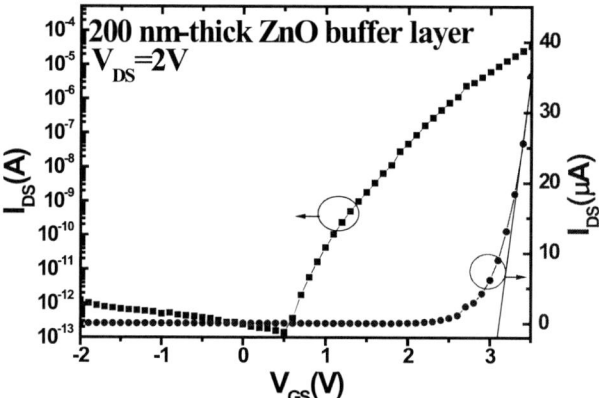

Figure 5. The I_{DS}-V_{GS} characteristics of ZnO:Al TTFTs with 200 nm-thick ZnO buffer layer.

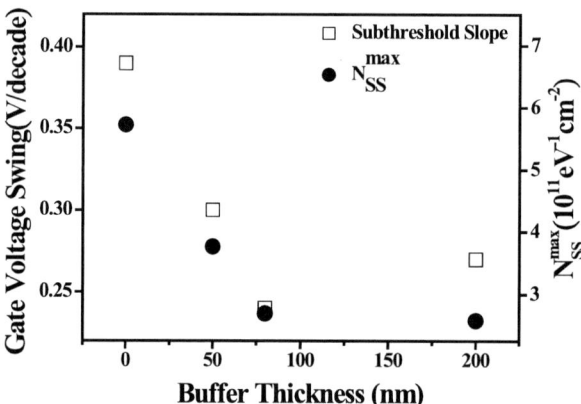

Figure 6. Dependence of the gate voltage swing and the maximum state density of ZnO:Al TTFTs on the thickness of ZnO buffer layer.

High responsivity ultraviolet photodetector based on *p*-GaN/*i*-ZnO nanorod /*n*-ZnO:In nanorod

Chia-Hsun Chen, Jheng-Tai Yan, and Ching-Ting Lee*

Institute of Microelectronics, Department of Electrical EngineeringNational Cheng Kung University, Tainan, Taiwan, Republic of China

Tel: 886-6-2082368   Fax: 886-6-2082368

E-mail: ctlee@ee.ncku.edu.tw

The *p-i-n* (*p*-GaN/*i*-ZnO nanorod/*n*-ZnO:In nanorod)-heterostructured nanorod photodetectors were fabricated using vapor cooling condensation method through anodic alumina membrane (AAM) plate. A conductive atomic force microscope system was employed to investigate the electrical properties of the single ZnO nanorod photodetector. The leakage current of the single *p-i-n*-heterostructured nanorod photodetector was lower than 5 pA for bias voltage up to 5V. The ultraviolet (360 nm)-visible (400 nm) rejection ratio was 66 and the photoresponsivity peaked at 360 nm was about 1461 A/W under reverse bias of -5V.

Introduction

Ultraviolet radiation detection has been attracted much attention for a wide range of traditional and emerging civil and military applications, such as chemical analysis, biological analysis, flame sensors, space-based optical communications. Recently, various types of ZnO-based photodetectors (PDs) have been reported, such as photoconductors, p-n junction diodes, p-i-n diodes, Schottky barrier photodetectors, metal-semiconductor-metal (MSM) photodetectors, phototransistor, photodetector array and avalanche photodetector. In this study, we used one-dimensional ZnO nanorod to fabricated single *p-i-n* heterostructured nanorod photodetector. The nanorod photodetedctor have good UV response to enhance the performance of UV detectors due to large surface area to volume ratio. Furthermore, individual ZnO nanostructure has very high internal photoconductivity gain due to the surface-enhanced electron-hole separation efficiency. Therefore, ZnO nanostructures become a promising material to make UV detectors.

Device structure and Fabrication process

The structure of a single *p*-GaN/*i*-ZnO nanorod/*n*-ZnO:In nanorod (*p-i-n*)-heterostructured nanorod photodetector is schematically shown in Fig. 1. A 800 nm-thick GaN buffer layer was grown on the *c*-plane (0001) sapphire substrates using a metal-organic chemical vapor deposition (MOCVD) system. A Mg-doped GaN (*p*-GaN) layer on was then grown on the buffer layer to work as the *p*-type layer of the single *p-i-n* heterostructured nanorod photodetectors. To perform the *p*-type ohmic contact, patterned Ni/Au (20/100 nm-thick) metals were deposited on the *p*-GaN layer and then annealed at 500°C for 10 min in an air ambient in the rapid thermal annealing system. In this study, the vapor cooling condensation system was used to grow *i*-ZnO nanorod and *n*-ZnO:In nanorod [1]. First, put the anodic alumina membrane (AAM) template on *p*-GaN substrate and fixed underneath stainless steel plate. The pore diameter and density of the AAM template used in this study are 200 nm and 5×10^9 pores/cm^2, respectively. In this system, ZnO powder was put on a tungsten boat and heated. The sublimated ZnO vapor gases were condensed and deposited on the p-GaN substrate which was cooled by liquid nitrogen. Finally, ZnO:In nanorods were deposited as the n-type ZnO nanorode to fabricated the single p-i-n heterostructured nanorod photodetector.

Experimental results and Discussion

The current-voltage characteristic of a single p-i-n heterostructured nanorod photodetector was measured by the nanoprobing system through a conductive atomic force microscope (CAFM) system. Figure 2 shows the measured current-voltage (I-V) characteristic. It can be seen that the leakage current of single p-i-n heterostructured nanorod photodetector was 4.9 pA for applied a reverse bias of -5V.

The spectral photoresponsivity of the fabricated p-i-n photodetector was measured in the wavelength from 340 nm to 420 nm using a 150W xenon (Xe) lamp, which was connected to a monochromator via an optical fiber. Each measured device was uniformly illuminated from the top side. Using the measured spectral photocurrents, the photoresponsivity spectra were calculated according to $R=I_{ph}/P_{opt}$, where R is the photoresponsivity, I_{ph} is the photocurrent and P_{opt} is optical power. The spectral photoresponsivity of the single p-i-n-heterostructured nanorod photodetector is shown in the inset of Fig. 2. The photoresponsivity of the detector biased at -5V was about 1461 A/W at 360 nm. The associated ultraviolet (360 nm)-visible (400 nm) rejection ratio was 66. The gain of the single p-i-n heterostructured nanorod photodetector could be estimated to be 5×10^3. Such a result indicated that an internal gain existed in the single p-i-n-heterostructured nanorod photodetector. This high internal gain could be attributed to the high density of hole-trap states at the nanorod surface.

As shown in Fig. 3(a), in the dark, excess oxygen was adsorbed by taking a free electron from the surface of the naonrod photodetector to form a chemically adsorbed surface state. This reaction could be expressed as:

$$O_2 + e^- \rightarrow O_2^- \qquad (1)$$

where e^- is electrons. Fig. 3(b) showed that, under illumination, electron-hole pairs are produced by light absorption when photon energy is higher than the fundamental band gap of ZnO. The adsorbed oxygen is photodesorbed from the surface:

$$h^+ + O_2^- \rightarrow O_2 \qquad (2)$$

where h^+ is holes. Therefore, photogenerated holes migrate to the surface and are trapped, leaving behind unpaired electrons in the nanorod that contribute to the photocurrent and increase the conductivity [2]. Thus, the single p-i-n-heterostructured nanorod photodetectors have a larger gain because of a larger ratio of surface area to volume, which can adsorb more excess oxygen at the nanorod surface.

Conclusions

The vapor cooling condensation method was used to grow single p-i-n-heterostructured nanorod photodetectors. The single p-i-n nanorod photodetector exhibited a low leakage current of 4.9 pA. The ultraviolet (360 nm)-visible (400 nm) rejection ratio was 6.6×10^1 under a reverse bias of -5V. The high photoresponsivity peaked at 360 nm was about 1461 A/W under a reverse bias of -5V, corresponding to gain of the single p-i-n heterostructured nanorod photodetector about be 5×10^3. The photosensitivity and photocurrent may come from photo induced chemidesorption of oxygen at the surface. The nanorod photodetector has a larger photocurrent because of the large surface area to volume ratio, which can adsorb more excess oxygen at the nanorod surface.

Acknowledgments

This work was supported from the National Science Council of Taiwan, Republic of China under Grant No. NSC 98-2119-M-006-005.

References

[1]   R. W. Chuang, R. X. Wu, L. W. Lai, and C. T. Lee, Appl. Phys. Lett. **91**, 231113 (2007).

[2]   C. Soci, A. Zhang, B. Xiang, S. A. Dayeh, D. P. R. Aplin, J. Park, X. Y.Bao, Y. H. Lo, and D. Wang, Nano Lett. **7**, 1003 (2007).

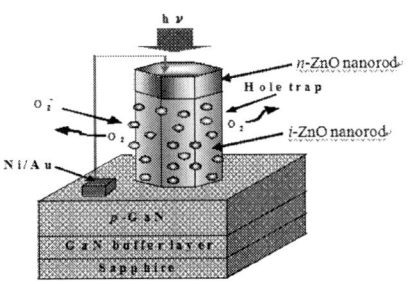

Fig. 1 Schematic of single *p-i-n* heterostructured nanorod photodetector.

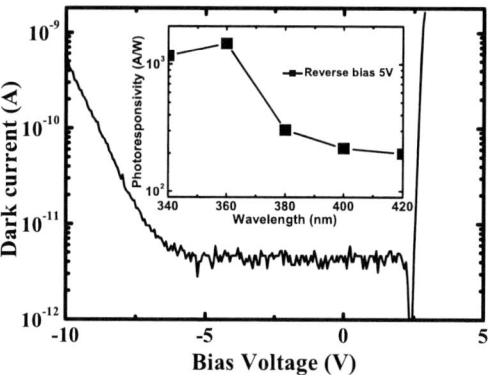

Fig. 2 Current-voltage characteristics of a single *p-i-n* heterostructured nanorod photodetector. The spectral photoresponsivity was shown in the inset.

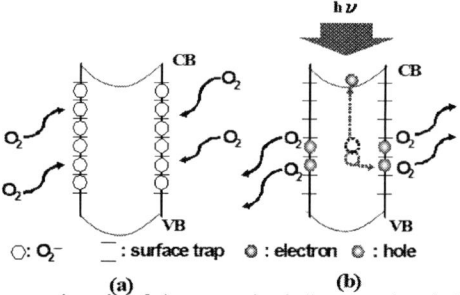

Fig. 3 Schematic of the energy band diagrams of a single *p-i-n* heterostructured nanorod photodetector: (a) in the dark and (b) under illumination.

Epitaxial Growth of ZnO on LiAlO$_2$ and LiGaO$_2$ Substrates by Chemical Vapor Deposition

Cheng-Huan Chen, Jun-Yi Yu, Teng-Hsing Huang, Liuwen Chang[*] and Mitch M.C. Chou

[a] Department of Materials and Optoelectronic Science/ Center for Nanoscience and Nanotechnology, National Sun Yat-Sen University, Kaohsiung 80424, Taiwan, R.O.C

> Epitaxial growth of ZnO on a tetragonal LiAlO$_2$ (100) substrate and an orthorhombic LiGaO$_2$ (100) substrate was examined using X-ray diffraction, scanning electron microscopy and transmission electron microscopy. Results indicated that the crystal structure plays a major role on the orientation selection. The m-plane ZnO is favorable to grow on both substrates due to its low lattice mismatch to the substrates. However, the kinetic factor cannot be ignored on epitaxial growth. At low temperatures, limited adatom mobility can lead to the nucleation of crystals having a less favorable orientation, such as the c-plane ZnO on LAO or the random oriented ZnO on LGO. In addition, the morphology of the epitaxially grown crystals is affected by the anisotropy of the growth rate. For the c-plane ZnO, high anisotropy at low growth temperature results in rod shape crystals. At higher temperatures, the anisotropy is decreased and lateral growth leads to wineglass shape crystals.

Introduction

Zinc oxide has drawn a great attention recently for possible applications on ultraviolet/blue semiconductor lasers and light emitting devices. Up to now, sapphire is the most popular substrate used for epitaxial growth of ZnO. However, the large lattice mismatch causes high density of threading dislocations in the epitaxial layer which deteriorates its electronic and optical properties [1-3]. γ-LiAlO$_2$ (LAO) and LiGaO$_2$ (LGO) substrates are attractive alternatives to replace sapphire as substrates for ZnO, since their lattice mismatches to ZnO are much lower than that of sapphire.

γ-LiAlO$_2$ has a tetragonal lattice with lattice parameters of a=0.5168 nm and c=0.6268 nm, whereas LiGaO$_2$ is orthorhombic with lattice parameters of a=0.5402 nm b=0.6372 nm and c=0.5007 nm [4, 5]. Figs. 1a and 1b show the atom arrangement of LAO and LGO, respectively, in the [100] direction. It is found that a tetragonal symmetry on the (100)$_{LAO}$ surface leads to a good match with the $(10\bar{1}0)$ of ZnO by aligning [010]$_{LAO}$ and [0001]$_{ZnO}$. The in-plane lattice mismatch in [010]$_{LAO}$//[0001]$_{ZnO}$ is 0.7% and that in [001]$_{LAO}$//[11$\bar{2}$0]$_{ZnO}$ is 3.6 %. In addition, a quasi-hexagonal symmetry of the anions and cations can be observed on the (100) $_{LAO}$ face. The lattice misfit in [010]$_{LAO}$//[10$\bar{1}$0]$_{ZnO}$ is rather large of 8.2%, which is still possible to lead to a c-plane, or (0001) oriented, epitaxial growth. In fact, both (0001) and $(10\bar{1}0)$ oriented ZnO epilayer have been reported to grow on the LAO (100) substrates [6, 7]. However, detailed study on the orientation selection mechanism is still absent. On the other hand, there is only a tetragonal anion/cation symmetry observed on the (100)$_{LGO}$ surface, as indicated in Fig.

1b. The in-plane lattice misfit in $[001]_{LGO}//[0001]_{ZnO}$ is 3.8% and that in $[010]_{LGO}//[11\bar{2}0]_{ZnO}$ is 2.7 %. Accordingly, m-plane, or $(10\bar{1}0)$ oriented, ZnO grown epitaxially on LGO substrate is expected, though has not been demonstrated yet.

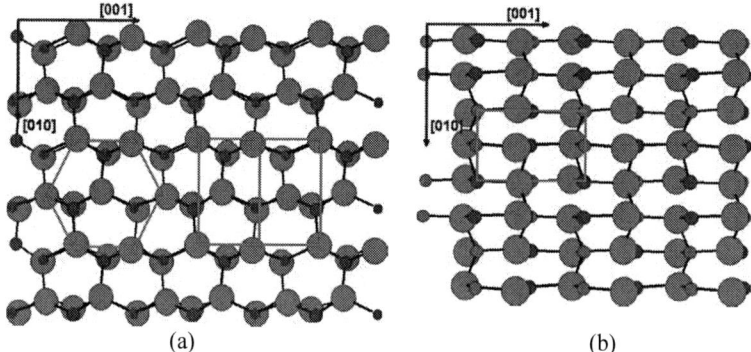

(a)   (b)

Fig. 1 Crystal structures of the (a) γ-LiAlO$_2$ and (b) LiGaO$_2$ phases in the [100] direction. Oxygen anions are solid red, lithium cations are solid blue, aluminum cations are solid purple and gallium cations are solid green.

In this work, the effects of substrate crystal structure and growth temperature on the orientation selection of ZnO deposited by chemical vapor deposition are studied. Results indicated that the crystal structure plays a major role on the orientation selection. However, the kinetic factor cannot be ignored on epitaxial growth. In addition, the morphology of the epitaxially grown crystals is affected by the anisotropy of the growth rate.

Experimental

γ-LiAlO$_2$ and LiGaO$_2$ single crystals of 2 inches in diameter were both grown in [100] direction using the Czochralski pulling technique. After the crystals were cut and polished, LAO and LGO (100) wafers were obtained and were used as substrates throughout the experiments. The substrates were cleaned in successive ultrasonic baths of acetone and methyl alcohol for 10 minutes each. The ZnO films were deposited on the wafers in a horizontal chemical vapor deposition (CVD) reactor having two temperature zones [6]. Zinc acetylacetonate hydrate $(Zn(C_5H_7O_2)_2 \cdot H_2O$, Lancaster, 98%) was used to be a zinc source and was loaded into the low-temperature zone of the reactor. The source temperature was controlled at 110 °C to vaporize the solid reactant. The vapor was carried by a N_2/O_2 mixture (500/500 sccm) flowing into the high temperature zone of the reactor in which the substrate was placed. Prior to the CVD process, the chamber was pumped to a pressure of 130 Pa or lower, and was then refilled with oxygen to maintain a constant pressure of either 1.3×10^4 or 2.7×10^4 Pa. The growth temperatures were in a range of 470~700 °C, and the deposition time was varied from 3000 seconds to 10,800 seconds.

Following the CVD growth, a field emission scanning electron microscope (SEM, JEOL 6330, operating at 5 kV) was used to examine the surface morphology of the epitaxial films. X-ray diffraction studies were performed either in a Siemens D-5000 diffractometer (Cu-Kα radiation) equipped with a graphite monochromater mounted in the secondary side or a Burker D8 diffractometer (Mo-Kα radiation) equipped with a Eularian cradle to monitor the orientation and crystallinity of the deposited films. Detailed microstructural characterization of the films was carried out by transmission electron microscopy (TEM, JEOL 3010 operating at 200kV). Cross-sectional TEM samples were prepared using the focus ion beam (SMI 3050) lift-out method.

Results and Discussion

ZnO on LAO substrate

Figs. 2a-d are the SEM secondary electron images of the ZnO deposited on LAO substrates at 650 °C (2.7×10^4 Pa), 570 °C (2.7×10^4 Pa), 570 °C (1.3×10^4 Pa) and 470 °C (2.7×10^4 Pa), respectively, for 10800 s. The substrates were all positioned in such a way that their [010] direction is parallel to the horizontal direction as indicated in Fig. 2a. At 650 °C, the entire substrate is covered by ZnO crystals of a long slate shape with a relatively low roughness (see Fig. 2a). EBSD analysis carried out in the previous study revealed that these slates have a ($10\bar{1}0$) azimuthal orientation with their [0001] direction parallel to [010]$_{LAO}$ [7]. It seems to indicate that the low lattice misfit allows the ZnO crystals to grow faster in the [0001]$_{ZnO}$ direction. Moreover, scattered crystals of a round shape can seldom be observed at the interactions of the slates. These ZnO crystals may have a (0001) azimuthal orientation as discussed below. Fig. 2b shows that the morphology of the ZnO crystals deposited at 570 °C is completely different from that observed at 650 °C. The film is composed of hexagonal-shape crystals having an average size of ~500 nm. These crystals are aligned in the same way with one of their hexagonal edges parallel to the [001]$_{LAO}$ direction. Previous EBSD result verified that these hexagonal crystals have a (0001) azimuthal orientation and possess an orientation relationship of the same as that predicted in Fig. 1a [7]. On decreasing the growth pressure from 2.7×10^4 Pa to 1.3×10^4 Pa, the resultant film has a similar morphology, but a larger size of ~1500 nm (see Fig. 2c). Finally, Fig. 2d shows that ZnO grows as thin hexagonal rods at 470 °C. Many rods have sizes of less than 100 nm, but some of them can be as large as ~250 nm. A close examination of the image shows that there are in fact two variants of rods co-existed. These two variants have a 30° rotational relationship in the [0001] direction, as shown in the circled area in Fig. 2d.

X-ray diffraction patterns acquired in an ω-2θ configuration for the ZnO samples deposited on LAO are shown in Fig. 3. It is interesting to note that ($10\bar{1}0$)$_{ZnO}$ peak is detected in all three samples, no matter what the crystal morphology is. The (0002) peak, however, only appears in the patterns for the 470 °C and 570 °C deposited samples. As a result, at temperatures below 650 °C, both the c-plane and the m-plane ZnO crystals can nucleate on LAO. The nucleation of the c-plane ZnO is, however, almost completely prohibited at 650 °C. In addition, the d-spacing of the ($10\bar{1}0$)$_{ZnO}$ plane increases slightly from 0.281 nm (475 °C and 575°C) to 0.282 nm at 650 °C, indicating that the cohesive strain in a compressive sense has probably not been relaxed completely yet. A very weak peak at 2θ=36.3° is observed in the 470 °C pattern. This peak is resulted from diffraction of the (002) plane of a β-LiAlO$_2$ phase which is found occasionally on the surface of the as-polished LAO wafer and the β phase transforms to the γ phase at temperatures higher than about 550°C [8].

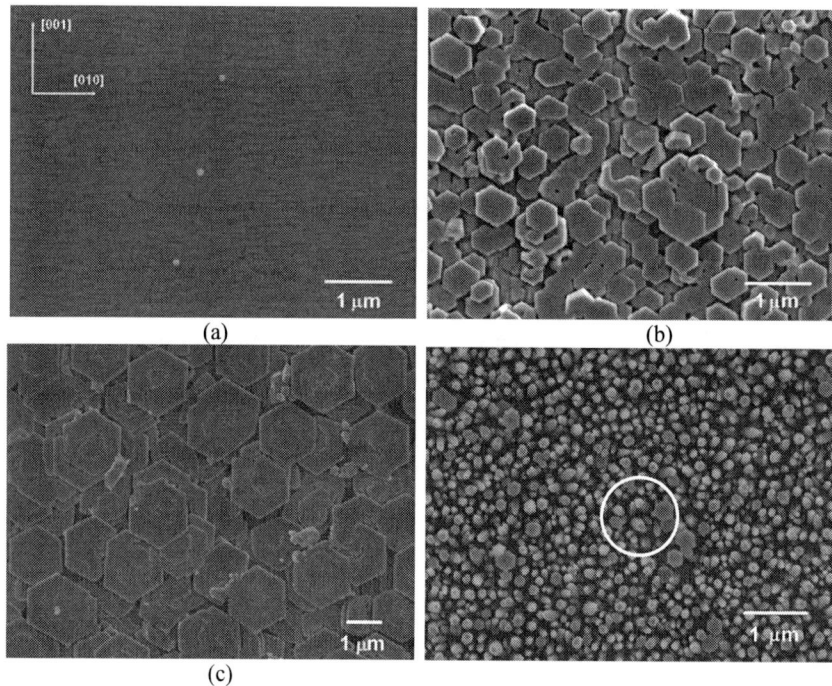

Fig. 2 SEM secondary electron images of the ZnO films deposited at (a) 650 °C (2.7×10^4 Pa), (b) 570 °C (2.7×10^4 Pa), (c) 570 °C (1.3×10^4 Pa), and (d) 470 °C (2.7×10^4 Pa).

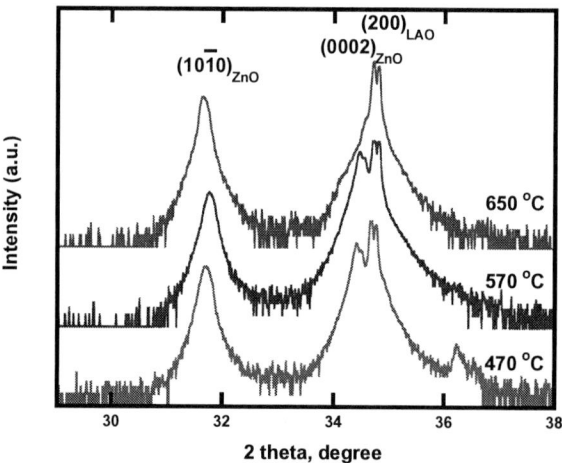

Fig. 3 X-ray diffraction patterns for the ZnO samples deposited at P=2.7×10^4 Pa.

Fig. 4a shows a cross section TEM bright field (BF) image of the sample deposited on LAO at 570 °C (2.7×10^4 Pa) for 3000 s. There are two kinds of crystals observed: dome-shaped crystals with a rounded top surface of 200-300 nm in height and wineglass-shaped ones with flat top surface of ~400 nm in height. The electron beam is parallel to $[001]_{LAO}$ (see Fig. 4b). The selected area diffraction (SAD) pattern of a wineglass shape crystal (labeled C in Fig. 4a) in Fig. 4c shows that the crystal has a [0001] azimuthal orientation. On the other hand, the SAD pattern shown in Fig. 4d indicates that the orientation of the dome shape crystal (labeled D in Fig. 4a) is orthogonal to that of the dome shape one about the $[001]_{LAO}$ axis. In other words, the former is a m-plane crystal. Both the c-plane and the m-plane crystals nucleate directly on the LAO substrate and possess well-defined orientation relationships with the substrate as suggested above. The in-plane growth rate of the m-plane ZnO is slightly lower than that of the c-plane one. In addition, the c-plane ZnO crystals start to grow laterally in the late stage of deposition and result in this wineglass shape. On prolonging the deposition time, the c-plane crystals grow laterally over the m-plane crystals and start to coalesce with each other as shown in Fig. 2b. The lateral growth is promoted when the sample is deposited at a lower pressure of

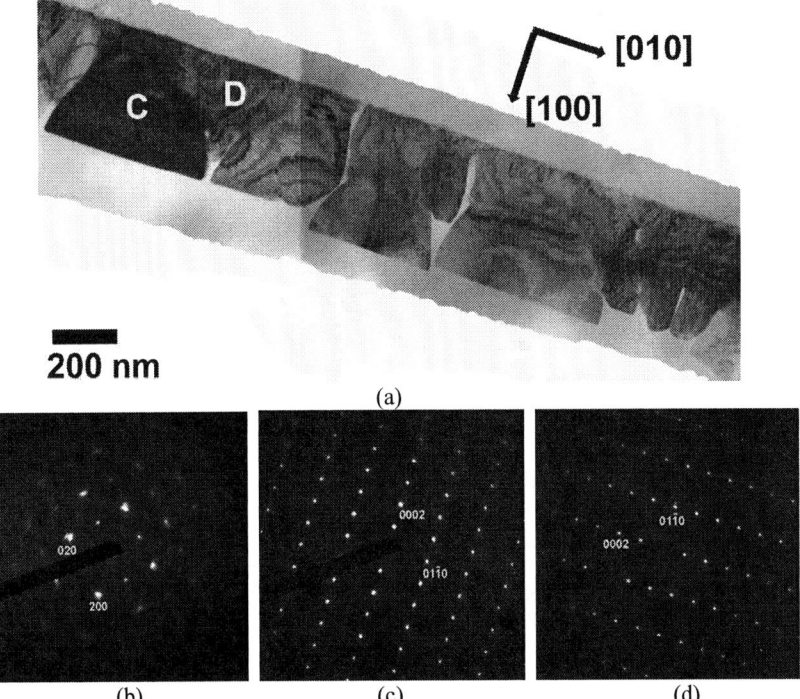

Fig. 4   Cross section TEM (a) bright field image, (b) SAD from LAO in [001] zone, (c) SAD from a crystal labeled C in [11$\bar{2}$0] zone, and (d) SAD from a crystal labeled D in [11$\bar{2}$0] zone for the film deposited at 570 °C for 3000 s.

1.3×10^4 Pa. Fig. 5 is a cross-section TEM photograph for a sample grown at 570 °C for 10800 s at a pressure of 1.3×10^4 Pa. The c-plane crystals first grow as rods of ~100 nm wide and then grow laterally into a wineglass or a up-side down hexagonal pyramid shape. Only one out of every ten to twenty rods can grow vertically to the surface and

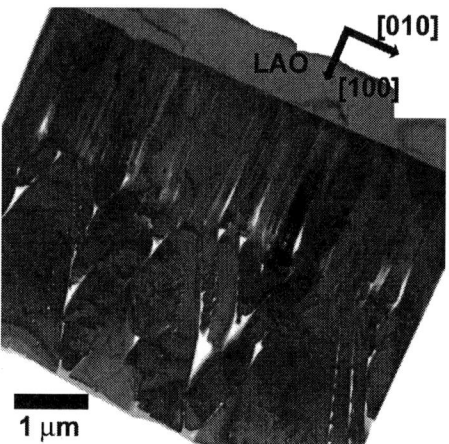

Fig. 5 Cross section TEM BF image of a sample grown at 570 °C for 10800 s at a pressure of 1.3×10^4 Pa

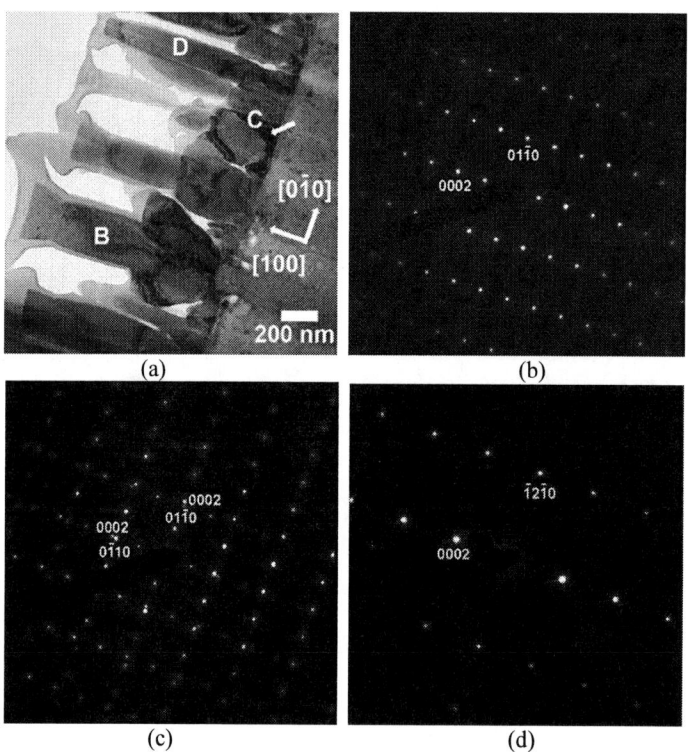

Fig. 6 Cross section TEM (a) bright field image, (b) SAD from a crystal labeled B in [$11\bar{2}0$] zone, (c) SAD from area C in two [$11\bar{2}0$] zones, and (d) SAD from a crystal labeled D in [$10\bar{1}0$] zone for the film deposited at 470 °C for 10800 s.

results in a large crystal size as observed in the in-plane direction. It is worth mentioning that the m-plane crystals are still observed at the ZnO/LAO interface in the present case.

Fig. 6a shows a cross section TEM bright field (BF) image of the film deposited on LAO at 470 °C for 10800 s. The electron beam is also parallel to the $[001]_{LAO}$ direction. Rods of 600~1200nm long and 50~250 nm wide are observed. There are also fine crystals of less than ~400 nm in height distributed between the rods. The TEM selected area diffraction patterns taken from a rod, labeled B in Fig. 6a, are shown in Fig. 6b. This rod is a c-plane crystal having its $[\bar{1}2\bar{1}0]$ direction parallel to $[001]_{LAO}$. Fig. 6c shows that diffraction patterns taken from area C consist two orthogonal $[\bar{1}2\bar{1}0]$ zone axis patterns. These two patterns are contributed from a c-plane nanorod and a fine crystal, indicated by an arrow in Fig. 6a, having its $[\bar{1}010]$ direction perpendicular to the substrate surface. In other words, the fine crystals distributed between the rods are the m-plane ZnO whose existence has been demonstrated in Fig. 3. The m-plane ZnO, however, probably possesses a low growth rate and its growth may be further shadowed by the fast growth of the (0001) oriented rods. Diffraction pattern of $[\bar{1}010]_{ZnO}$ zone axis shown in Fig. 6d, taken from a rod labeled D in Fig. 6a, verifies the existence of the 30° rotated c-plane crystal. The orientation relationship between the rod and substrate changes to $(0001)_{ZnO}//(100)_{LAO}$ and $[\bar{1}010]_{ZnO}//[001]_{LAO}$.

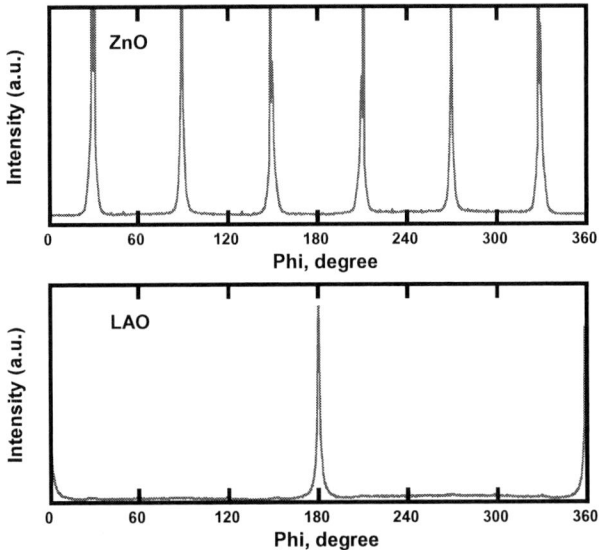

Fig. 7 X-ray phi scan for the $(302)_{LAO}$ plane and the $(10\bar{1}2)_{ZnO}$ plane for the sample deposited at 470 °C for 10800 s.

The formation of 30° domains in epitaxial growth of c-plane ZnO and GaN on sapphire substrate has been well documented [9-10]. It is attributed to the co-existence of a three-fold Al sublattice and a six-fold O sublattice on the (0001) sapphire surface [10]. However, the atomic arrangement on the LAO (100) surface is very different from that on the sapphire (0001) surface. The existence of the 30° rotation domains in c-plane ZnO deposited on LAO substrate may be attributed to a coincidence lattice-like arrangement since both $d(10\bar{1}0)_{ZnO}/d(001)_{LAO}$ and $d(11\bar{2}0)_{ZnO}/d(010)_{LAO}$ are very close to 1/3 [11]. Phi scans for the $(302)_{LAO}$ and the $(10\bar{1}2)_{ZnO}$ diffractions were carried out for the sample deposited at 470 °C, and results are shown in Fig. 7. The phi scan result shows six peaks of $(10\bar{1}2)_{ZnO}$ corresponding to the $[11\bar{2}0]_{ZnO}//[001]_{LAO}$ orientation relationship but no peaks at 30° offset are observed, indicating that these 30° domains are in vary low fraction.

ZnO on LGO substrate

Figs. 8a-c are the SEM secondary electron images of the ZnO films deposited at various temperatures for 7200 s on the LGO substrates. Fig. 8a reveals that the film deposited at 550 °C is composed of two kinds of ZnO crystals. One is randomly oriented ZnO crystals of 100~400 nm in size and another has a long slate shape. The slates are ~100 nm wide in $[010]_{LGO}$ and 200~1000 nm long in $[001]_{LGO}$. As mentioned before that the lattice mismatch in $[001]_{LGO}//[0001]_{ZnO}$ is 3.8% and that in $[010]_{LGO}//[11\bar{2}0]_{ZnO}$ is 2.7 %. The shape of the slates should be close to a squire if the misfit strain is the dominant factor determining the morphology of the crystals. However, the long slates shown in Fig. 8a demonstrate that either the cohesive strain has not been fully relaxed or the kinetic factors, such as the anisotropy of the growth rate, also affect the shape of the crystals. The film deposited at a higher temperature of 650 °C is almost all covered by ZnO slates with still high aspect ratio (see Fig. 8b). When the deposition temperature was increased to 700 °C, the morphology of the crystals changed slightly as shown in Fig. 8c. The size of the slates increases of about two to three times and well defined facets are developed. In addition, there are some areas which do not covered by the ZnO crystals yet after 3000 s deposition. A close examination shows that these areas are actually covered by a layer of nanocrystals. Cross section SEM observation (result not shown) showed that the nanocrystals covered a great portion of the substrate and the m-plane crystals grew laterally over these nanocrystals. Fig. 8d shows that a smooth film is obtained for prolonging the deposition time to 10800 s.

Fig. 9 shows the XRD patterns for the films deposited at different temperatures. At 550 °C, the pattern contains a relatively strong $(10\bar{1}0)_{ZnO}$ peak and two weak peaks of $(0002)_{ZnO}$ and $(10\bar{1}1)_{ZnO}$. The strong $(10\bar{1}0)_{ZnO}$ peak is mainly contributed by the aligned slate-shaped crystals whereas the weak peaks are from the randomly oriented crystals shown in Fig. 8a. No other peaks than the $(10\bar{1}0)_{ZnO}$ and $(200)_{LGO}$ peaks were detected for the film deposited at 650 °C, coinciding with the SEM observation that the film is mainly composed of the m-plane ZnO. In other words, the nucleation rate of the m-plane crystals relative to that of randomly oriented crystals is increased dramatically with increasing deposition temperature. At 700 °C, the pattern also contains one ZnO peak and one substrate peak within the two-theta range from 29° to 38°. However, very weak peaks at 43.5° and 57.5° were observed in a scan of a wider range (result not shown). These nanocystals therefore may be a $ZnGa_2O_4$ phase. It has been reported that a nanoscale $ZnGa_2O_4$ layer is formed at the interface of GaN and ZnO as well as GaAs and ZnO due to the inter-diffusion of Zn and Ga at temperatures of 700 °C or higher [12, 13]. Further characterization of the nanocrystal layer is, indeed, needed and is in progress.

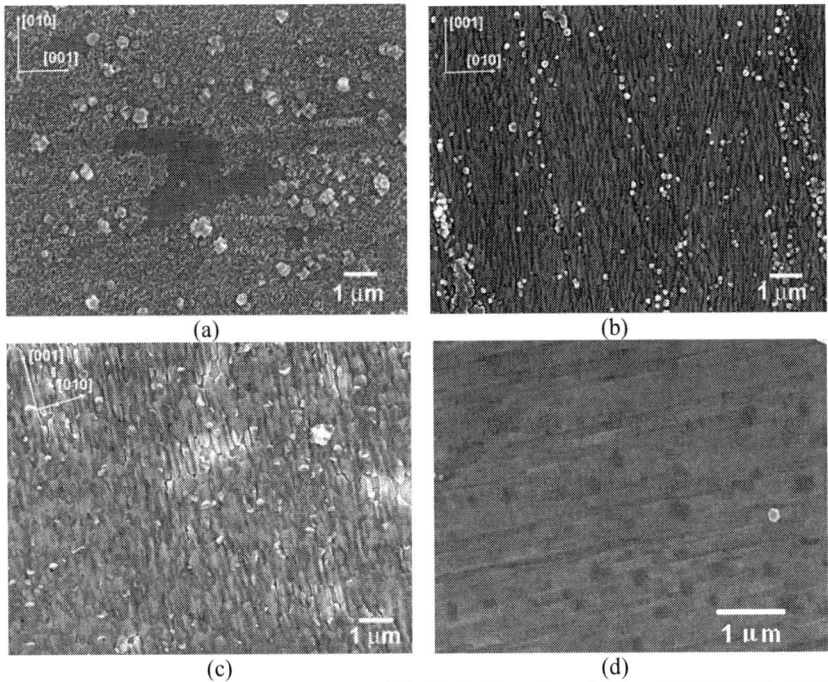

(a)   (b)

(c)   (d)

Fig. 8 SEM secondary electron images of the ZnO films deposited at (a) 550 °C for 7200 s, (b) 650 °C for 7200 s, (c) 700°C for 3000s and (d) 700 °C for 10800s.

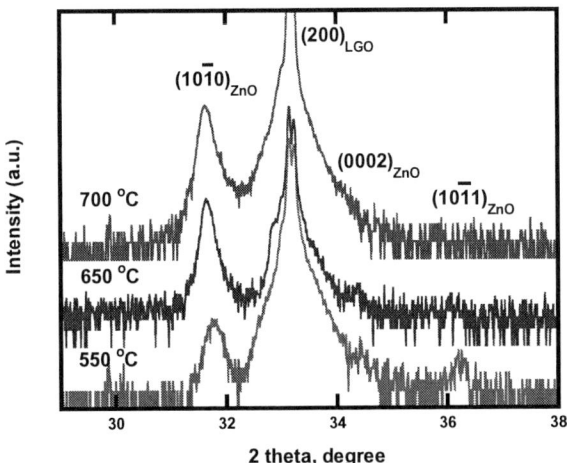

Fig. 9 X-ray diffraction patterns for the ZnO films deposited on LGO

In order to verify the orientation relationship between ZnO and LGO, x-ray phi scans were carried out for both ZnO and LGO. Fig. 10 shows the x-ray phi scan results for the

$(120)_{LGO}$ plane and the $(11\bar{2}0)_{ZnO}$ plane of a sample deposited at 700 °C. The result shows that the orientation relationship between ZnO and LGO can be described as: $[11\bar{2}0]_{ZnO}//[010]_{LGO}, (10\bar{1}0)_{ZnO}//(100)_{LGO}$, coinciding with the prediction in Fig. 1b.

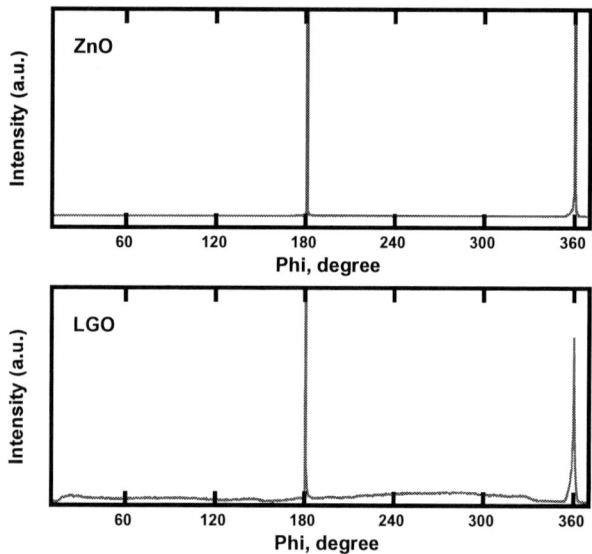

Fig. 10 X-ray phi scans for $(120)_{LGO}$ plane and the $(11\bar{2}0)_{ZnO}$ plane for the sample deposited at 700 °C for 3000 s.

The results clearly demonstrate that the orientation selection of ZnO, deposited by chemical vapor deposition, is affected by the substrate crystal structure and the growth temperature. Both the c-plane and the m-plane ZnO can grow epitaxially on the LAO (100) substrate due to the coexistence of a tetragonal site and a quasi-hexagonal site. On the other hand, the epitaxial growth of the c-plane ZnO on the LGO (100) substrate is totally prohibited for the lack of available nucleation site. In addition, kinetics may play an important role on nucleation. At high growth temperatures of 650 °C or above, the adatoms have enough mobility to reach sites having the lowest energy. As a consequence, only the m-plane ZnO can nucleat on LAO and LGO. However, at lower temperatures, the kinetic limitation makes epitaxial growth of the c-plane ZnO on LAO to be possible. For the LGO substrate, low adatom mobility results in the nucleation of randomly oriented ZnO. It is worth noting that the nucleation of the m-plane ZnO cannot be completely prohibited even at temperature as low as 470 °C. A mixture of the c-plane and the m-plane ZnO is obtained at low substrate temperatures.

The anisotropy of the growth rate of ZnO is the main reason corresponding to different morphologies of the resultant crystals. The growth rate in c axis of ZnO is higher than those in directions perpendicular to the c axis. Fig. 6a showed that the length of the c-plane rods is about three times of the height of the m-plane crystals at 475 °C. This high growth rate in $[0001]_{ZnO}$ direction results in a hexagonal rod with six {10-10} side faces as shown in Fig. 11a. At a higher temperature of 575 °C, the anisotropy of growth rate is reduced and therefore the c-plane crystals grow laterally into a wineglass or a up-side down hexagonal pyramidal shape as shown in Fig. 11b. As mentioned above, the growth rate anisotropy may also be responsible for the morphology of the m-plane crystals which is always elongated in c-direction no matter how large the lattice mismatch is. However, the confinement of the lattice matching plays a role on the morphology development. The m-plane crystals therefore developed a sward blade-shape when the confinement is completely released.

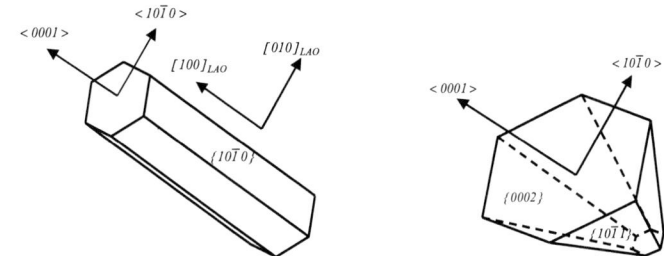

Fig. 11 Schematic drawing of the morphology of the c-plane ZnO grown at (a) 470 °C and.(b) 570 °C on LAO.

It is interested to note that a self-masked overgrowth phenomenon is observed for ZnO grown on LAO and LGO. For the LAO substrate, the c-plane ZnO can overgrow the m-plane one and form a continuous and smooth layer. On the other hand, the possible $ZnGa_2O_4$ nanocrystals formed on LGO at 700 °C also play a role to mask the substrate. The m-plane ZnO can therefore grow laterally over the nanocrystals and form a smooth layer. The potential benefit of the overgrowth is that an epitaxial layer with low density of threading dislocations can be obtained.

Conclusions

In conclusion, the m-plane ZnO are favorable to grow on both substrates due to its low lattice mismatch to the substrates. However, the kinetic factor cannot be ignored on epitaxial growth. At low temperatures, limited adatom mobility can lead to the nucleation of crystals having a less favorable orientation, such as the c-plane ZnO on LAO or the random oriented ZnO on LGO. In addition, the morphology of the epitaxially grown crystals is affected by the anisotropy of the growth rate. For the c-plane ZnO, high anisotropy at low growth temperature results in a rod shape crystals. At higher temperatures, the anisotropy is decreased and lateral growth leads to a wineglass shape crystal.

Acknowledgments

This work was financially supported by the National Science Council, R.O.C. under grant number NSC97-2221-E-110-009 and NSC98-2221-E-110-039-MY2, and by the Center for Nanoscience and Nanotechnology, National Sun Yat-Sen University.

References

1. S. J. Rosner, E. C. Carr, M. J. Ludowise, G. Girolami, and H. I. Erikson, *Appl. Phys. Lett.*, **70**, 420 (1997).

2. T. Hino, S. Tomiya, T. Miyajima, K. Yanahima, S. Hashimoto, and M. Ikeda, *Appl. Phys. Lett.*, **76**, 3421 (2000).
3. J. Y. Shi, L. P. Yu, Y. Z. Wang, G. Y. Zhang, and H. Zhang, *Appl. Phys. Lett.*, **80**, 2293 (2002).
4. M. Marezio, *Acta Cryst.*, **19**, 396 (1965).
5. M. Marezio, *Acta Cryst.*, **18**, 481 (1965).
6. M. M. C. Chou, L. Chang. H.-Y. Chung, T.-H. Huang, J.-J. Wu, and C.-W. Chen, *J. Cryst. Growth*, **308**, 412 (2007).
7. L. Chang, M. M. C. Chou, D.-S. Hwang, C.-W.Chen, *Phys. Status Solidi A*, **206**, 215 (2009).
8. R. R. Vanfleet, J. A. Simmons, D. W. Hill, M. M. C. Chou, and B. H. Chai, *J. Appl. Phys.*, **104**, 093530 (2008).
9. X. L. Du, M. Murakami, H. Iwaki, Y. Ishitani, and A. Yoshikawa, *Jpn. J. Appl. Phys., Part 2*, **41**, L1043 (2002).
10. K. Nakaharaa, H. Takasua, P. Fonsa, K. Iwataa, A. Yamada, K. Matsubaraa, R. Hungera, S. Niki, *J. Cryst. Growth*, **227-228**, 923 (2001).
11. A. Trampert, K. H. Ploog, *Cryst. Res. Technol*, **35**, 793 (2000).
12. S.-H. Hwang, T.-H. Chung and B.-T. Lee, *Mater. Sci. Eng. B*, **157**, 32 (2009)
13. H. F. Liu, A. S. W. Wong, G. X. Hu and H. Gong, *J. Cryst. Growth*, **310**, 4305 (2008).

CHAPTER 2

III-NITRIDES

46

Quasi-Ballistic Hole Transport in an AlGaN/GaN Nanowire

M.A. Mastro[a], H.-Y. Kim[b], J. Ahn[b], J. Kim[b], J. Hite[a], C.R. Eddy, Jr[a]

[a] U.S. Naval Research Laboratory, Washington, D.C. 20375, USA
[b] Department of Chemical and Biological Engineering, Korea University, Seoul, South Korea

> An AlGaN/GaN nanowire, with an isosceles-triangle cross-section, was designed to create a large negative polarization at the (000-1) facet. The resultant band bending at this interface formed a two-dimensional potential well that accumulated a hole gas. Transistor operation based on the two-dimensional hole gas showed characteristics of quasi-ballistic transport. A small number of elastic scattering sites were evident from quantum interference characteristics in the current-voltage data.

Introduction

Transport in current state-of-the-art transistors is no longer adequately described by classical drift-diffusion. Current silicon-based transistor technology, with an approximate 32nm channel length, exhibits carrier transport at approximately 50% of the ballistic limit, which is referred to the quasi-ballistic regime (1). The onset of true ballistic transport in Si(Ge) nanotransistors is at a channel length of approximately 10nm. Comparatively, transport in systems of reduced lateral dimension, such as a carbon nanotubes and AlGaAs/GaAs nanowire transistors, have shown ballistic transport behavior at channel lengths of hundreds of nanometers (2). This article discusses the formation of a hole gas in an AlGaN/GaN nanowire and the manifestation of conductance fluctuations that are distinctive of transport in the quasi-ballistic regime.

Calculations

Polarization

The III-nitride high electron mobility transistor is enabled by the large positive spontaneous and piezoelectric polarization at the AlGaN/GaN interface, which creates an enormous polarization-induced electron gas without the necessity to intentionally introduce dopants. Polarization in a III-nitride multi-layer nanowire is more complex as the polarization fields are dependent on the orientation of each facet (3,4). Growth of III-nitride nanowires via a VLS mechanism tends to proceed in the <11-20> a-plane direction with an isosceles-triangle cross-section. Depending on the growth conditions, two distinct sets of facets can form: (0001), (1-10-1), (-110-1); or (000-1), (-1101), (1-101). The total polarization at the (0001) AlGaN/GaN facet is similar to a standard thin film HEMT; in contrast, (000-1) AlGaN/GaN will posses a negative polarization and, depending on the band structure, the semiconductor will respond to these fields by accumulating holes.

The spontaneous polarization results from a discontinuity in the metal/N atom repeating structure that forms along c-axis. Therefore, the spontaneous polarization is proportional to the cosine of the angle of the semi-polar axis relative to the natural c-axis (5). The angle-dependence of the piezoelectric field is more complex as the strain relative to the semi-polar growth direction (figure 1) is first transformed to the natural c-axis orientation before calculating the tensor-product of the piezoelectric tensor and the strain (figure 2). The strain and resultant polarization fields at the semi-polar facets display a complex relationship to alloy composition, thickness, and piezoelectric coefficients.

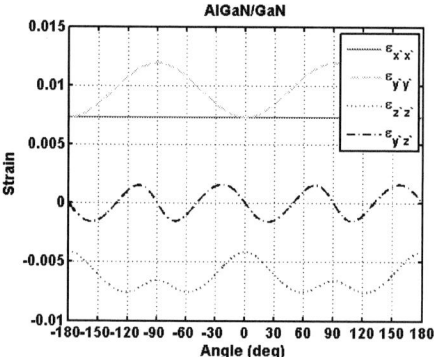

Figure 1. Elastic strain components for a thin $Al_{0.3}Ga_{0.7}N$ layer on semi-polar GaN. The angle represents the inclination of the semi-polar surface normal away from the natural c-axis of the crystal (4).

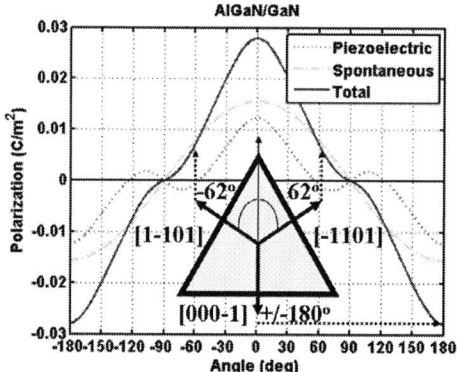

Figure 2. (a) Calculated piezoelectric, spontaneous, and total polarization fields for a pseudomorphic $Al_{0.3}Ga_{0.7}N$ layer on semi-polar GaN using piezoelectric coefficients given in Vurgaftman and Meyer (6). Insert: schematic cross-section of an AlGaN/GaN shell/core nanowire with the <11-20> normal to the page (4).

Carrier Density

Control of the polarization fields in this AlGaN/GaN nanowire allows the construction of nanowire transistor based on polarization-induced hole carriers. A coupled Schrodinger-Poisson calculation was conducted to account for quantum sub-bands and polarization fields in the nanowire. The total polarization ($P=P_{sp}+P_{pz}$) incorporates into the Poisson equation as

$$\nabla \cdot D = \nabla \cdot (\varepsilon E + P) = \rho,$$ [1]

$$\nabla \cdot [(\varepsilon(-\nabla\phi)] + \nabla \cdot P = \rho, \qquad [2]$$

assuming $\nabla \varepsilon \to 0$ then

$$\nabla^2 \phi = -\frac{\rho}{\varepsilon} + \frac{1}{\varepsilon}[\nabla \cdot P], \qquad [3]$$

where the polarization occurring at hetero-interface results in a fixed polarization charge, ρ^{Pol}, to give

$$\nabla^2 \phi = -\frac{\rho}{\varepsilon} - \frac{\rho^{Pol}}{\varepsilon}. \qquad [4]$$

The charge, ρ, in the semiconductor is primarily composed of hole carriers, p, electron carriers, n, ionized donors, N_D^+, and ionized acceptors N_A^- as well as donor and acceptor traps (7). Substituting these factors gives a functional form of the Poisson equation as

$$\nabla^2 \phi = -\frac{q}{\varepsilon}[N_D^+ - N_A^- + p - n] - \frac{\rho^{Pol}}{\varepsilon}, \qquad [5]$$

which describes the formation of charged carriers in semiconductor in response to the potential field. As can be seen in figure 3, the polarization field creates an large density of hole carriers on the GaN side of the (000-1) $Al_{0.3}Ga_{0.7}N$ / GaN interface even without intentional doping.

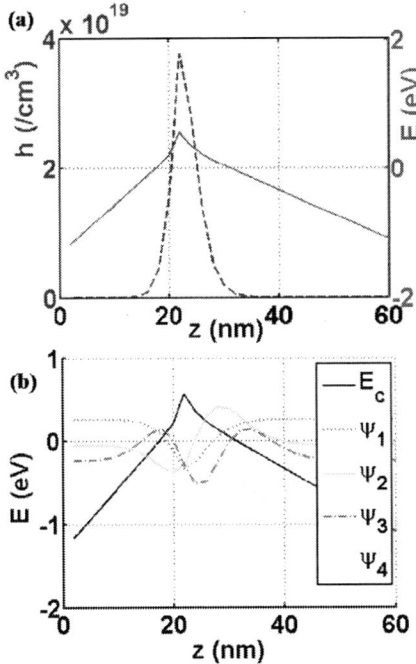

Figure 3. Calculated (a) carrier concentration (dashed line) and conduction band (solid line), and (b) conduction band (solid line) and the first four subbands for a 22nm $Al_{0.3}Ga_{0.7}N$ / GaN structure.

Carrier Transport

The drift-diffusion description of carrier transport begins to break down as the length, L, of the conduction channel approaches the average distance between scattering events. If the distance traveled is less than a mean free path between scattering events then the electron will accelerate similar to transport in a vacuum tube (8). In general, the scattering mechanisms can be classified as either elastic or inelastic collisions. The interaction between electrons and fixed impurities are elastic as there is minimal energy transfer. Conversely, inelastic scattering arises from the collision between electrons and electrons, or electrons and quantized lattice vibrations (phonons) (9). Electrons experiencing an elastic collision will maintain its phase memory while an inelastic collision will cause an electron to lose its phase coherence. Thus it beneficial to define an average mean free length between elastic collisions, L_m, and a mean free length between inelastic collisions, L_φ.

A model was developed to describe transport along a 500nm nanowire length via the hole gas at the (000-1) AlGaN/GaN interface. The entire geometry of the wire was modeled; however, the carriers were mainly confined in a 2DHG defined by the potential well and the width of the (000-1) GaN facet. The non equilibrium Green function approach, as described by Lundstrom and Guo (10) and Datta (11), allows the introduction of small number of random elastic and inelastic scattering sites in the conduction channel. Figure 4 shows that a small ensemble of elastic scattering sites creates conductance fluctuations in the current-voltage relationship. A quantum interference pattern forms from the interaction of the carrier wavefront with a set of elastic scattering centers. This particular interference pattern is native to this nanowire and is reproducible for an exact applied voltage (or magnetic) field (12).

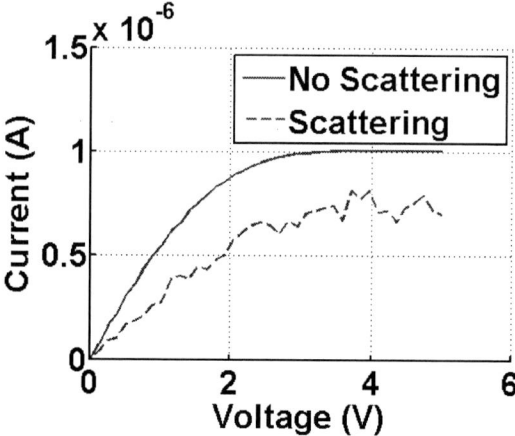

Figure 4. Calculated current-voltage behavior of the 2DHG nano-transistor with a channel length approaching the elastic phase conserving length and the inelastic phase distorting length. The effect of a small number of elastic scattering sites is to create a diffraction pattern of fluctuations (12). In contrast, the inelastic scattering sites dampen these conductance fluctuations as well as the overall current through the transistor.

Experimental

A 0.05 M nickel nitrate solution was repeatedly dripped onto a silicon substrate and blown dry in N_2 then loaded into a vertical impinging flow, metal organic chemical vapor deposition reactor. The GaN nanowire core was grown at a temperature of 850°C and the AlGaN shell was grown at 950°C. The higher growth temperature favored formation of nanowires with a (000-1) facet. The III-nitride nanowires were dispersed in 50 ml iso-propyl alcohol solution by sonication, and subsequently dispersed onto a 300nm SiO_2 / Si wafer. The SiO_2/Si wafer was pre-patterned with finger-shaped Ti/Au electrodes (20 nm/80 nm thickness). FIB deposited Pt contact lines from the source (or drain) region of the nanowire to large area contact pads. A Ni/Au layer was deposited on the backside of the conductive wafer to provide a gate contact for the NW-FET.

Current-voltage characteristics of such an undoped AlGaN/GaN p-type nanowire transistor are presented in figure 5. Conductance fluctuations seen in figure 5 are similar to the fluctuations predicted above in figure 4 for a theoretical nanowire with an ensemble of elastic scattering sites. These diffracting objects predicatively introduce conductive fluctuations in the output waveform. This time-independent effect is not related to time-dependent noise such as Johnson noise, shot noise, and 1/f noise (12).

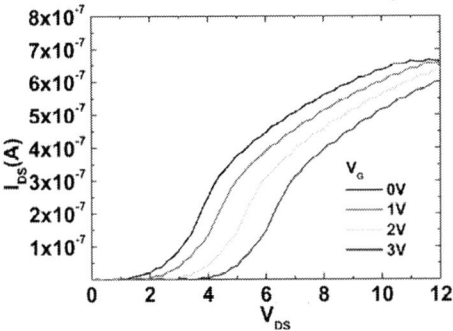

Figure 5. Current-voltage behavior of the p-type AlGaN/GaN core / shell nanowire transistor with polarization induced hole carriers at the (000-1) interface.

Acknowledgments

Research at the US Naval Research Lab is partially supported by the Office of Naval Research.

References

1. R. Chau, S. Datta, M. Doczy, B. Doyle, J. Kavalieros, M. Metz, IEEE Elec. Dev. Lett., **25**, 408 (2004)
2. R. Chau, J. Brask, S. Datta, G. Dewey, M. Doczy, B. Doyle, J. Kavalieros, B. Jin, M. Metz, A. Majumdar, M. Radosavljevic, Microelectronic Eng., **80**, 1 (2005)
3. M.A. Mastro, B. Simpkins, M. Twigg, M. Tadjer, R.T. Holm, C.R. Eddy, Jr., ECS Trans., **13**-3, 21 (2008)

4. M.A. Mastro, B. Simpkins, G.T. Wang, J. Hite, C.R. Eddy, Jr., H.-Y. Kim, J. Ahn, J. Kim, Nanotechnology, In Press (2010)
5. S. Vandenbrouck, K. Madjour, D. Theron, Y.J. Dong, Y. Li, C.M. Lieber, C. Gaquiere, IEEE Electron Dev. Lett., **30**, 322 (2009)
6. I. Vurgaftman, J. Meyer, J. Appl. Phys., **94**, 3675 (2003)
7. J.C. Freeman, NASA technical report: Basic Equations for the Modeling of Gallium Nitride High Electron Mobility Transistors, NASA/TM 211983 (2003)
8. M.P. Anantram, M.S. Lundstrom, D.E. Nikonov, Cond. Mat. 0610247 (2007)
9. G.W. Hanson, Fundamentals of Nanoelectronics, Prentice Hall (2007)
10. M.S. Lundstrom, J. Guo, Nanoscale Transport: Device Physics, Modeling, and Simulation, Springer (2005)
11. S. Datta, Quantum Transport: Atom to Transistor, Cambridge University Press (2005)
12. D.F. Holcomb, Am. J. Phys. **67**(4): 278 (1999)

Light Extraction Enhancement of n-side up Light-Emitting Diodes without
Electrodes Covering by Wafer Bonding and Textured Surfaces

R. H. Horng[a], Y. A. Lu[b] and D. S. Wuu[c]

[a] Department of Electro-Optical Engineering, National Cheng Kung University, Tainan
701, Taiwan, ROC
[b] Institute of Precision Engineering, National Chung Hsing University, Taichung 402,
Taiwan, ROC
[c] Department of Science and Materials Engineering, National Chung Hsing University,
Taichung 402, Taiwan, ROC

Textured n-side-up GaN LED with interdigitated imbedded electrodes (IIE)
eliminating the electrode-shading loss with high reflection mirror and
double-side roughening on both p-GaN and undoped-GaN layers have been
investigated. The epitaxial layers of the device are grown on (0001)
sapphire substrates by metal-organic chemical vapor deposition. The
devices are subsequently fabricated with wafer-bonding, laser lift-off,
chemical dry/wet etching techniques. This n-side up structure was useful to
enhance light extraction and increase the light output power. We compared
the performance of the luminance intensity(at 350 mA injection current),
which is 78.85 % and 115.38 % higher than those of the conventional
structure and the p-side up structure with high reflection mirror on silicon
substrate with electrode shading, respectively.

Introduction

Wide bandgap light-emitting diodes (LEDs) that are nitride-based, ranging from
ultraviolet to the short-wavelength part of the visible spectrum have been developed in
recent years [1]. These devices make new display technology such as traffic signals, LCD-
TV, and backlight for cell phones have become feasible[2]. Due to the lack of native
substrates, films of GaN and related nitride compounds are commonly grown on sapphire
wafers. The conventional LEDs are not efficient because the photons are emitted in all
directions. A large fraction of light emitted were limited in the sapphire substrate, could
not contribute to usable light output. Moreover, the poor thermal-conductivity of the
sapphire substrate is also an problem associated with conventional nitride LEDs. Therefore,
free-standing GaN optoelectronics without sapphire are most desirable. The substrate
transfer technique is now a well-known progress in achieving high brightness LEDs. Thin-
film laser lift-off (LLO) technique[3] combine with n-GaN surface roughening has
generally been established as an effective tool for Nitride-based heteroepitaxial structures
to eliminate the sapphire constraint [4,5], the structure is regarded as a good candidate for
enhancing the light extraction efficiency of GaN-based LEDs. The high reflection mirror
reflector on silicon substrate is one of the techniques for the optical confinement as well.
However, there still exists the electrode shading problem. The emitting light will be cover
and absorb because of the electrodes and the light efficiency consequently will reduce. In
this paper, n-side-up devices with interdigitated imbedded electrodes (IIE) for overcoming
the electrode-shading problem are provided and fabricated.

Experiment

The GaN LED epitaxial layers of the conventional structure, proven to be of high quality [6], are grown on c-plane (0001) sapphire substrates by metal-organic chemical vapor deposition. Each chip with an area of 1×1 mm^2 is fabricated. The electrode contact pads are patterned with Cr/Au (25 nm/200 nm) by thermal evaporation. The schematic cross section of the conventional structure is shown in Fig. 1(a). Undoped-GaN and indium tin oxide (ITO) act as an epitaxial buffer layer, and a transparent conducting layer for current spreading, respectively. The process flow of the GaN/InGaN LEDs with the imbedded interdigital electrodes structure is as following. First, the top side (p-side) of the conventional device is upside down bonded onto a thin-film-coated silicon substrate, which consists of SiO$_2$ (900 nm) and Ti/Al (15 nm/ 250 nm), using epoxy glue bonding. Laser lift-off process is then employed for sapphire removal. Subsequently, the pattern of the electrode contact pads is defined by photolithography, which is followed by a highly anisotropic etching process, using an inductively coupled plasma reactive ion etching (ICP-RIE) to expose p- and n-contact pads with an etching depth of 6 μm and 5.3 μm, respectively. The imbedded interdigital electrodes device is consequently finished with the wet processing of H$_3$PO$_4$ and NaOH solution for reducing the thickness and roughening the surface of the n-GaN window layer, respectively, as shown in Fig. 1(c). In order to compare the performance, a conventional LED with sapphire substrate and p-side up thin film LEDs were prepared, and so is the structure with the high reflection mirror reflector from thin-film approach which is shown in Fig. 1(b). All LEDs are fabricated using the same epi-wafer and standard photolithography and dry etch techniques, characterized under the same condition.

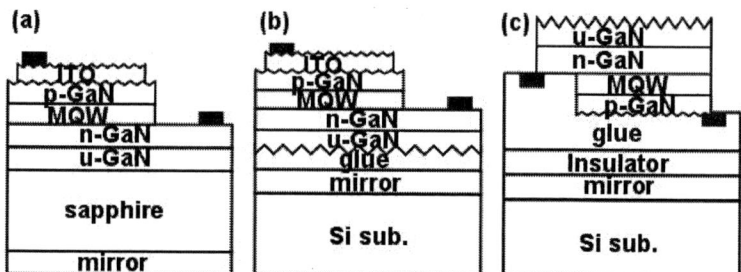

Fig. 1 Cross-sectional of (a) conventional structure, (b) p-side up GB-LED on silicon substrate with high reflection mirror, (c) n-side up GB-LED on silicon substrate, with an imbedded interdigitated electrode structure.

Results and Discussion

Figs. 2 show the optical photograph of these three type devices. Fig. 2(a) is the conventional LEDs with sapphire substrate. Figs. 2(b) and (c) are the thin-film p-side up and n-side up LEDs, respectively. It is worthy to mention that the same LEDs epilayers were transferred to the same Si substrate with the same mirror. However, the electrodes are on the top surface for the p-side up thin-film LEDs. Correspondingly, there is no electrode shading for the n-side up thin-film LEDs due to the electrodes being imbedded under the epilayer. It can be further confirmed by the secondary electron microscope (SEM) as shown in the Fig. 2 (d). On the other hand, the epilayers present milky color for the thin film LEDs regardless of p-side up or n-side up thin film LEDs. It is attributed to the epilayers with double sides surface roughening and mirror substrate. As the white light incidents into the epilayer, the light will be scattered by these roughening surface or reflected by the bottom mirror and then scattering by the surfaces.

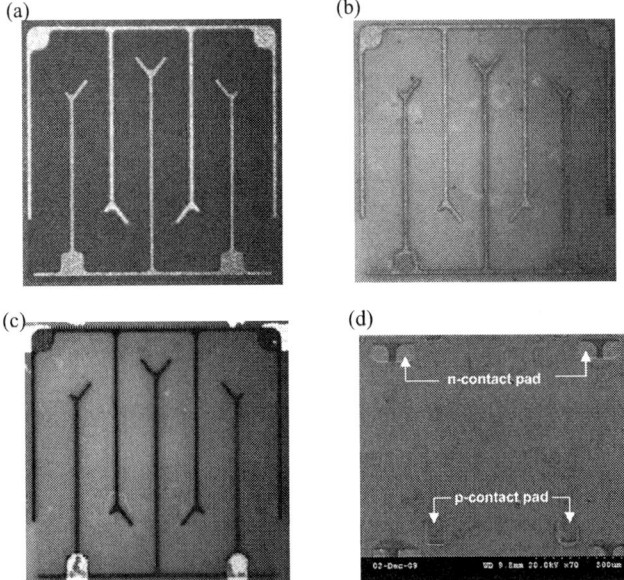

Fig. 2 Optical photograph of (a) conventional LEDs with sapphire substrate, (b) thin film p-side up LED with mirror deposited Si substrate and (c) thin film n-side up LED with mirror deposited Si substrate. (d) Microphotography observed by SEM of the thin film n-side up LED.

Because the fabricated thin-film LEDs must undergo several processes, such as LLO, glue bonding, and n-side roughening, it is important to evaluate whether these processes destroy the LED characteristics or not. Fig. 3 presents the current and resistance as the function of voltage (I-V, Rs-V) for the thin film LEDs and compared with the original GaN/sapphire LED. It was found that the I-V curves of original GaN/sapphire LED, p- and n-side up glue-bond LEDs are almost the same. This suggests these processes could not destroy the electrical properties of the thin-film LEDs. The forward voltages (@ 20 mA) of the original, p- and n-side up glue-bond LEDs are 2.60, 2.63 V and 2.62 V, respectively. The resistance has presented no difference among these three samples. It is because the device fabricated before the epilayer transferring processes. The almost same resistance for these three samples suggest that the roughening process of n-side surface does not effect on the resistance. On the other hand, the reverse leakage current (not shown)of the original, p- and n-side up glue-bond LED structures are all below -2 μA at -5 V. These results imply that the electrical characteristics of the LED devices are not destroyed during the fabrication processes described earlier.

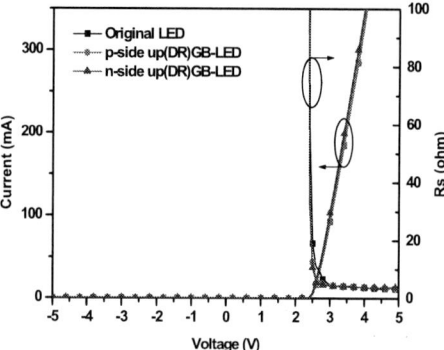

Fig.3 Current and Resistance as functions of voltage for the original, p-side-up thin film and n-side-up thin film LEDs.

The corresponding characteristics of luminance intensity verse current are shown in Fig. 4 (without lens encapsulated). It was found that n-side up glue-bonded LED shows the highest light output performance as compared with that of conventional LED and p-side up glue-bonded LED. An improvement of 115.38 % for the luminance intensity from 104.1 mcd (original LED) to 224.1 mcd (n-side up GB-LED) at 350 mA is achieved. Noted that the performance of p-side up GB-LED(at 350 mA injection current) is only 78.85 % higher than the conventional structure. The light extraction efficiency of n-side up thin film LEDs is better than that of the p-side up thin film LEDs. The enhancement could be attributed to especially the eliminating of the electrode-shading loss, the effect of double-side roughening, and high reflection mirror. Therefore, this effect plus double-side roughening provides the generated photons more probabilities to escape from the active region. It is worthy to note that the area of electrode shading is about 13 % of the light emitting area. However, there is about 20.4% enhancement of the brightness as compared the n-side up thin film LEDs with p-side up thin film LEDs. Obviously, the chemical surface roughening could be the other contributor to improving light extraction efficiency.

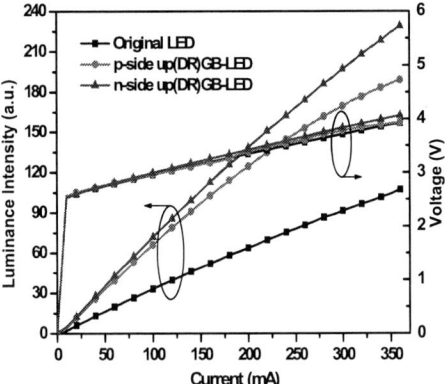

Fig. 4 Luminance intensity of original, p-side-up thin film and n-side-up thin film LEDs as a function of injection current.

From the Figs. 4, we can safely conclude that even though the p-side up GB-LED can enhance the light extraction, the effect of light extraction on the device is not as good as that of the n-side up GB-LED transferred onto high reflection mirror. The formation of random roughening on the u-GaN could contribute to the luminance intensity improvement via significantly enabling the photons emitted from active region to randomize effectively and redirecting photons back into the escape cone through the semiconductor to air interface resulting in additional light extraction.

As concerning this point, it can be confirmed by the cross-sectional images (observed by SEM)of the three types LED and shown in Figs. 5(a), (b), and (c). The structure of p- and n-side up GB-LED in Fig. 5, corresponding to Fig. 1(b) and 1(c), can be clearly observed and contrasted for each layer except the undistinguishable boundary of p-QW-n layers. The intact surface of the window layer and the thickness of the epitaxial layers (~7.5 μm) can be seen in Fig. 5(a) right after the up-side-down pattern transfer and the sapphire removal. After the chemical etching time of 3 minutes and the surface roughening, a 5-μm epitaxial layer with a pyramidal textured surface is shown in Fig. 5(c). As a result, the pyramidal textured surface regions are supposed to cause the undesired reflection and deteriorate the optical confinement. Therefore, the evidence from Fig. 5 shows that the mechanism of the optimized light extraction efficiency may not only electrode shading, but also the pyramidal textured surface.

Fig. 6 Cross-sectional SEM images of the (a) original (single roughness), (b) p-side up (double roughness) and (c) n-side up (double roughness) structure LEDs.

Additionally, the surface temperatures were measured for the LEDs, shown in Figs. 5. We observed that as the forward current increases over 350 mA, the current spreading performance of n-side up GB-LED is the best one, better than that of p-side up GB-LED as well. The results reveal that the current spreading at larger drive current could be attributed to a good thermal conductivity of Silicon carrier.

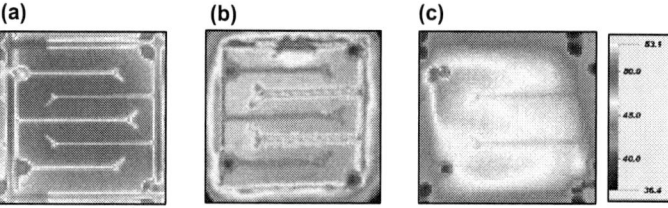

Fig. 5 Surface temperatures of (a) original, (b) p-side up (c) n-side up structure LEDs.

CONCLUSION

In this paper, the n-side up structure was useful to avoid light-absorbing and enhance the light efficiency. The GaN/InGaN double-heterojunction LED, in the absence of electrode-shading loss, are demonstrated with the improved luminance intensity of 224.1 mcd (at 350 mA injection current) which is 115.38 % higher than those of the conventional structure. For p-side up GB-LED, which is 78.85 % higher than the conventional structure, yet worse than n-side up GB-LED. It reveals that the improvement of light efficiency is attributed to the light extraction including the surface texture, the high reflection mirror reflector, and the imbedded electrodes structure.

Acknowledgments

This work was supported by the National Science Council, by the Ministry of Economic Affair and the Ministry of Education (Taipei, Taiwan) under contract no.

NSC95-2221-E-005-147, 97-EC-17-A-07-SI-097 and Aiming for the Top University plane, respectively.

References

1. E. F. Schubert and J. K. Kim, *Science*, **308**, 1274 (2005).
2. A. Zukauskas, M. S. Shur, and R. Gaska, *Introduction to Solid-State Lighting*. New York: Wiley.
3. Z. S. Luo, Y. Cho, V. Loryuenyong, T. Sands, N. W. Cheung, and M. C. Yoo, *IEEE Photon Technol. Lett.* **14**, 1400 (2002).
4. M. K. Kelly, O. Ambacher, B. Dahlheimer, G. Groos, R. Dimitrov, H. Angerer and M. Stutzmann, *Appl. Phys. Lett.*, **69**, 1749 (1996).
5. W. S. Wong, T. Sands and N. W. Cheung, *Appl. Phys. Lett.*, **72**, 599 (1998).
6. X. Zheng, R. H. Horng, D. S. Wuu, M. T. Chu, W. Y. Liao, M. H. Wu, R. M. Lin, and Y. C. Lu, *Appl. Phys. Lett.*, **93**, 261108-1(2008).

AlGaN/GaN HEMT Based Biosensor

Siddharth Alur[1], Tony Gnanaprakasa[2], Yaqi Wang[1], Yogesh Sharma[1], Jing Dai[2], Jong Wook Hong[2], Aleksandr L. Simonian[2], Michael J. Bozack[1], Claude Ahyi[1] and Minseo Park[1]*

[1]Department of Physics, Auburn University, Auburn, Alabama 36849
[2]Materials Research and Education Center, Department of Mechanical Engineering, Auburn University, Auburn, Alabama 36849
*corresponding author

GaN has been considered as a promising candidate for biosensing due to its chemical and thermal stability. In this work, an AlGaN/GaN high electron mobility transistor (HEMT) wafer with a 2DEG mobility of 1300 cm^2/v-s and a sheet carrier density of 1×10^{13} cm^{-2} was used as a sensor platform. Ti/Al/Ni/Au were used as source and drain contacts, with Ni/Au contacts as gate electrodes. A NF$_3$ plasma was used for device isolation. Photodefinable PDMS was used for the purpose of encapsulation. Once the encapsulation was completed using the photodefinable PDMS, the Schottky contact was exposed to a thiolated DNA in immobilization buffer for a period of 12hours. XPS was used to confirm the probe immobilization. The change in the (I$_d$-V$_{ds}$) characteristics of the device is measured during these processes, which confirm the probe immobilization and also the probe and the target DNA hybridization.

Introduction

Semiconductor based biosensors for detecting DNA sequences is an area of intense research due to its applications in biotechnology, molecular biology and the pharmaceutical industry. Several different methods including optical (1-2), changes in mass (3), electrochemical detection (4-5) have been developed for this purpose. Most of these methods include using a labeling agent (6) or are expensive, time consuming, tedious and destructive. Semiconductor biosensors based are better than the above-mentioned procedures since they are label free, nondestructive and fast (7-10).

In semiconductor biosensors a field effect transistor (FET) can be used in converting a biological signal into an electrical signal. AlGaN/GaN based high electron mobility transistors (HEMT) are ideally suited for such devices since they have a high electron sheet concentration induced by spontaneous and piezoelectric polarization of the strained AlGaN and the GaN layer (11-12). Kang and others have demonstrated the successful detection of DNA using the AlGaN/GaN field effect transistor (13).

In this work, we use a 22 base pair best representing salmonella DNA to show the functioning of the device as a DNA hybridization sensor. DNA hybridization is the process by which the complementary sequences of 2 SS-DNA bind to form a DS-DNA. The DNA molecules are negatively charged in an aqueous solution. When a SS-DNA hybridizes with another SS-DNA, which has a matching complementary sequence, there is a change in the charge density on the sensor surface on the device where this hybridization has occurred. This change in the charge density will result as a change in the current voltage characteristics of the device.

Experiment

The HEMT structure consists of a thin AlN layer, 2.7μm GaN buffer, 20nm AlGaN and a 2nm GaN cap layer. The HEMT wafer was purchased from SVTA. The GaN cap layer helps in reducing the surface states and in improving the ohmic source and drain contacts without having any ill effects on the schottky contact. The epilayers were grown on top of sapphire. The 2DEG mobility is greater than $1300cm^2/v$-s at room temperature. Ti/Al/Ni/Au were deposited and then annealed at 750°C for 30s under flowing N_2 gas which made the ohmic contacts. These processes were performed by the standard lithographic, sputtering and lift-off techniques. The Schottky contacts were made by sputtering Ni/Au (14-15). The I-V characteristics of this device for varying gate biases are measured (figI). After making the gold wire bonding to the ohmic contacts photodefinable polydimethylsiloxane (PDMS) was spin coated on the device. This spin coated photodefianble PDMS was then patterned by lithographic techniques so that we have a 20μm thick encapsulation layer covering the device except for the gate area of each device.

The DNA immobilization on the gold surface is performed using the following procedure: The exposed gold-coated gate area is first cleaned with methanol and then acetone. It is then further cleaned with a solution mixture containing H_2O_2, NH_4OH and DI H_2O present in the ratio of 1:1:5. After washing the surface thoroughly with DI H_2O the surface is then treated with 50μl of thiolated probes (1μM) 5'-SH-$(CH_2)_6$-GGTGGTGCTAAGGCAATGATAG-3' in the immobilization buffer for 12hours. The surface is then washed with immobilization buffer and then treated with 50μl of 1mM MCH in Abs. ethanol for 90mins. The I-V characteristics of the device is then measured after these procedures. The DNA immobilized surface is then exposed to the target SS-DNA in the hybridization buffer solution. The I-V characteristics of the device are then measured after waiting for a period of about 45mins and the stability of the device is verified. The change in the I-V characteristics of the device during these processes is measured. XPS is used to confirm the probe immobilization.

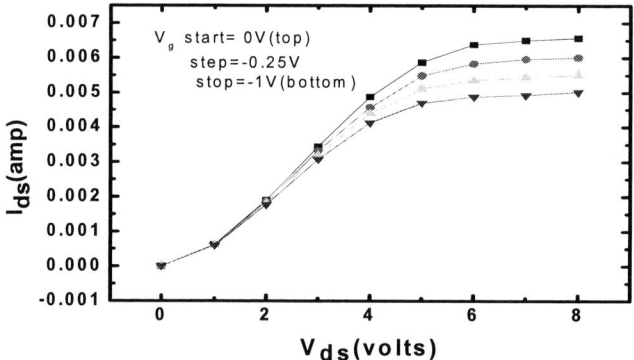

Figure I. I-V characteristics of the HEMT for varying gate voltage

Results and Discussion

The change in the (I_d-V_{ds}) characteristics of the device is measured during these processes (Figure II). The change in the current voltage characteristics of the device confirms the probe immobilization and also the probe and the target DNA hybridization. Since we are not using a reference electrode superimposed with a high frequency ac signal the device takes longer time to stabilize after exposure to the buffer solutions. A high resolution XPS scan over the sulphur S2P peak shows a wide peak with a peak binding energy of about 162 eV (Figure III). Such a binding energy has been observed in similar, well-characterized chemical systems bonded to Au. These results have to be repeated to be checked for reproducibility and stability of the device.

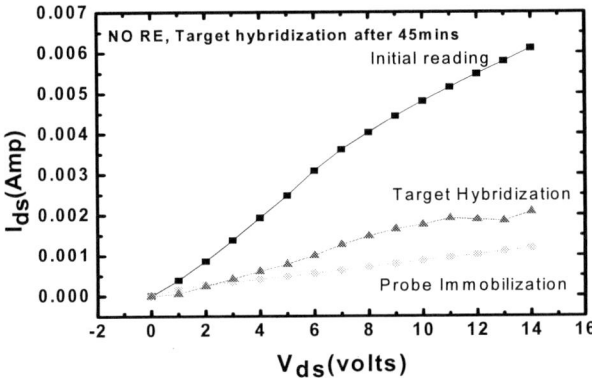

Figure II. Change in the I-V characteristics after probe immobilization and after target hybridization

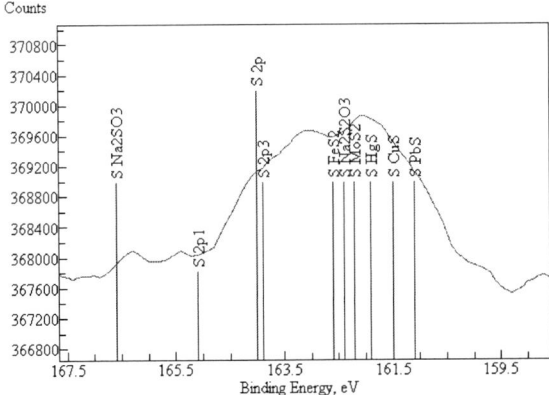

Figure III. High resolution XPS scan of the S2P peak

Acknowledgments

The project was funded by USDA through AUDFS.

References

1. Pease, A. C. Solas, D. Sullivan, E. J. Cronin, M. T. Holmes, C. P. Fodor, S. P. A. Proc. Natl. Acad. Sci. USA **91**, 5022 (1994).
2. P. A. E. Piunno, U. J. Krull, R. H. E. Hudson, M. J. Damha, H. Cohen, Analytica Chimica Acta **288**, 205 (1994).
3. Y. Okahata, M. Kawase, K. Niikura, F. Ohtake, H. Furusawa, Y. Ebara, Analytical Chemistry **10**, 1288 (1998).
4. Rajesh, T. Ahuja and D. Kumar, Sens. Actuat. B: Chem. **136** , 275 (2009).
5. E. Palecek and M. Fojta, Talanta **74**, 276 (2007).
6. T. A. Taton, C. A. Mirkin, and R. L. Letsinger, Science **289**, 1757 (2000).
7. E. Souteyrand, J. P. Cloarec, J. R. Martin, C. Wilson, I. Lawrence, S. Mikkelsen, and M. F. Lawrence, J. Phys. Chem. B **101**, 2980 (1997).
8. J. Fritz, E. B. Cooper, S. G. Audet, P. K. Sorger, and S. R. Manails, Proc. Natl. Acad. Sci. U. S. A. **99**, 14142 (2002).
9. F. Pouthas, C. Gentil, D. Cote, and U. Bockelmann, Appl. Phys. Lett. **84**, 1594 (2004).
10. G. Xuan, J. Kolodzey, V. Kapoor, G. Gonye, and Appl. Phys. Lett. **87**, 103903 (2005).
11. S. J. Pearton, B. S. Kang, S. Kim, F. Ren, B. P. Gila, C. R. Abernathy, J. Lin, and S. N. G. Chu, J. Phys.: Condens. Matter **16**, R 961 (2004).
12. B. S. Kang, F. Ren, L. Wang, C. Lofton, Weihong Tan, S. J. Pearton, A. Dabiran, A. Osinsky, and P. P. Chow, Appl. Phys. Lett. **87**, 023508 (2005).
13. B. S. Kang, S. J. Pearton, J. J. Chen, F. Ren, J. W. Johnson, R. J. Therrien, P. Rajagopal, J. C. Roberts, E. L. Piner, and K. J. Linthicum, Mater. Res. Soc. Symp. Proc. Vol. **955**, 0955-114-06 (2007).
14. Y. Wu, B. Keller, P. Fini, S. Keller, T. Jenkins, L. Kehias, S. DenBaars, and U. Mishra, IEEE Electron Device Lett. **19**, 50 (1998).
15. Y. F. Wu, S. Keller, P. Kozodoy, B. P. Keller, P. Parikh, D. Kapolnek, S. DenBaars, and U. K. Mishra, IEEE Electron Device Lett. **18**, 290 (1997)

ECS Transactions, 28 (4) 65-70 (2010)
10.1149/1.3377101 ©The Electrochemical Society

Development of Enhancement Mode AlN/Ultrathin AlGaN/GaN HEMTs by Selective Wet Etching

T. J. Anderson[a], M. J. Tadjer[b], M. A. Mastro[a], J. K. Hite[a], K. D. Hobart[a], C. R. Eddy, Jr[a], and F. J. Kub[a]

[a] Power Electronics Branch, Naval Research Laboratory, Washington, DC 20375 USA
[b] Department of Electrical and Computer Engineering, University of Maryland, College Park, MD 20742, USA

A novel recessed-gate structure involving an ultrathin AlGaN barrier layer capped by an AlN layer in the source-drain access regions has been implemented to demonstrate enhancement-mode high electron mobility transistors. A wet etch process has been developed using heated photoresist developer to selectively etch the AlN for the gate recess step, bypassing plasma etching and associated issues. The etch has been proven to be selective to AlN over AlGaN and GaN. A repeatable threshold voltage of +0.21V has been demonstrated with 4 nm AlGaN barrier layer thickness.

Introduction

There is currently significant interest in developing AlGaN/GaN High Electron Mobility Transistors (HEMTs) for both microwave and power switching applications due to their high electron mobility relative to Si, and higher sheet carrier density and higher breakdown voltage than GaAs or Si. The material system has demonstrated excellent thermal and chemical stability, as well as high mobility and breakdown field. This device typically operates in depletion mode, but enhancement mode operation is highly desirable particularly for power switching or digital logic circuit applications. Most efforts toward normally-off operation to date have focused on processing methods to remove charge under the gate by either etching the AlGaN to below the critical thickness to deplete the channel by plasma etching in the gate region[1-3], or by implanting negative charge in the AlGaN layer under the gate by exposing the device to a fluorine-based plasma[4-6]. While selective wet etching is desirable to bypass plasma processing, and is commonly used in the GaAs and InP industries, no etch stop structure has been demonstrated to date for the GaN/AlGaN system.

This work follows up on a previous demonstration of enhancement mode operation with a selective wet etch structure[7]. This structure involved a GaN buffer layer, followed by an ultrathin (<8 nm) AlGaN barrier, the thickness of which determined the threshold voltage, capped with an AlN layer to maintain a high strain-induced 2DEG. Wet etching of AlN using heated AZ400K photoresist developer has been reported in the literature[8]. Experiments have been performed to verify that this process is selective to AlGaN as well as GaN, and this structure has been implemented to fabricate functional HEMTs. Using this structure, an ultrathin AlGaN layer can be used for threshold voltage control, taking advantage of the precision thickness control from MOCVD. The polarization charge provided by the thin AlGaN alone is not sufficient to support high sheet charge in the channel, thus an AlN cap is required to increase the polarization charge and reduce resistance in the source-drain access regions.

65

Experimental

AlN/AlGaN/GaN layer structures were grown on a-plane Al_2O_3 substrates by MOCVD. The layer structure included an initial 2μm undoped GaN layer on an AlN nucleation layer, followed by either 8 or 4 nm $Al_{0.3}Ga_{0.7}N$ layers, followed by 4 nm AlN cap layer grown at 1050 °C. The sheet resistance for both samples was ~1100 Ω/sq, measured with a Lehighton probe and confirmed by TLM. The sheet carrier concentration was ~6×10^{12} cm^{-2} with a mobility of ~ 1000 cm^2/V-s at room temperature for both samples, determined by Hall measurements. Mesas were formed using an Inductively Coupled Plasma etch with Cl_2/Ar chemistry. Ohmic contacts were then deposited by lift-off of e-beam evaporated Ti/Al/Ni/Au, followed by an RTA at 900C for 30s in flowing N_2 atmosphere. Ohmic contact was made through the AlN layer, with a resulting contact resistance of 2×10^{-6} Ω-cm^2. To form a hard mask for the gate opening etch, PECVD SiN_X was deposited and subsequently patterned using a SF_6-based RIE etch, also opening the ohmic contacts. The gate opening process implemented selective chemical etching of the AlN using AZ400K developer at 85 °C. The gate was then formed by lift-off of e-beam evaporated Ni/Au. A schematic of the device structure is shown in Figure 1.

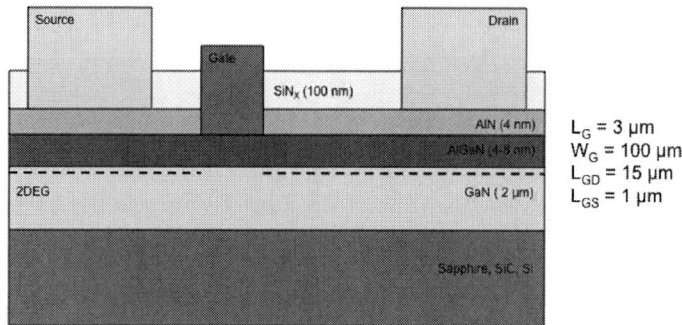

Figure 1. Cross section diagram of AlN/ultrathin AlGaN/GaN HEMT

To test etch selectivity, the wafer with 4 nm AlN/8 nm AlGaN/2 μm GaN was diced into 8 pieces and etched with durations from 0-70 minutes in 10 min intervals. It is clear from scanning electron microscopy (SEM) on ungated samples (not shown) that the gate dimension becomes larger after 20 min etching compared to the as-patterned feature in the hard mask. This implies that the etch mechanism is isotropic, as it etched laterally in the source-drain region and visibly undercut the SiNx mask for the very long etch times. The open gated current was measured before and after etching. The devices were then gated for FET I-V measurements, discussed below. Further characterization using X-ray photoelectron spectroscopy (XPS) has confirmed the selectivity of the etch.

Results

After depositing gates on the devices from the 4 nm AlN/8 nm AlGaN/2 μm GaN wafer, the FET I-V curves were measured and the threshold voltage was extracted as a function of etch time, shown in Figure 2. There is clearly a shift from -1.3V initially to -

0.4V after 10 minutes etching, then remaining constant for the remainder of the etch time studied. The constant threshold voltage implies that the AlGaN barrier is intact, and its thickness does not change with etch time. Also shown in Figure 2 is the on-resistance of the devices, which increases initially, due to the high-resistance region under the gate being formed when the channel becomes depleted, then remaining constant up to 20 min, then further increasing as the lateral etch broadens the feature and increases the resistance in the access regions. This clearly supports the etch stop hypothesis. The current level continues to degrade with etch time as the AlN is laterally etched, thus decreasing polarization and increasing the resistance in the access regions.

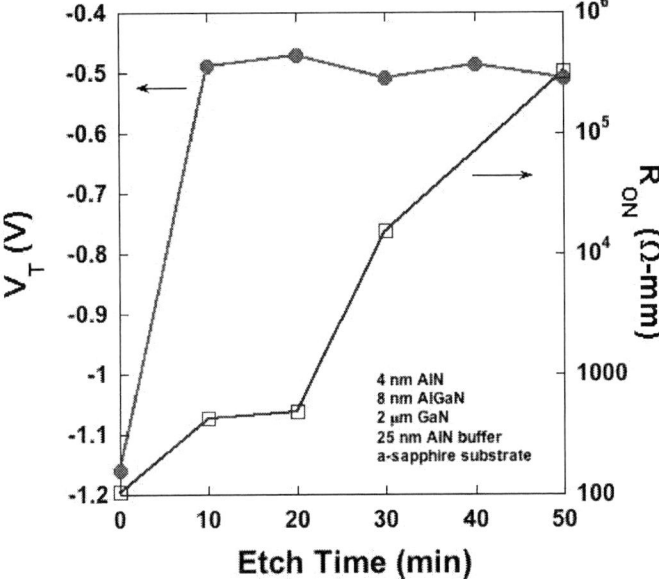

Figure 2. Threshold voltage and on-resistance as a function of etch time.

Based on the initial positive results, a wafer was grown with a 4 nm thick AlGaN layer, which was expected to yield a positive threshold voltage based on fundamental calculations described below. Fabrication was completed as described previously. From the FET I-V curves, shown in Figures 3 and 4, a threshold voltage of +0.21 V was extracted. This value was reproducible within 0.02V across the entire quarter wafer that was processed. The saturation current was 41 mA/mm at Vg= 2.5V and the maximum transconductance was 33 mS/mm at Vg = 0.5V. While current density was low due to the relatively high sheet resistance and large gate length (3-5 µm) and gate-drain gap (15-20 µm), it is comparable to that found in an unetched structure with similar sheet resistance. The mesa-to-mesa isolation current was 0.2 mA at 10V, which implies that off-state current is limited by buffer leakage for this particular sample.

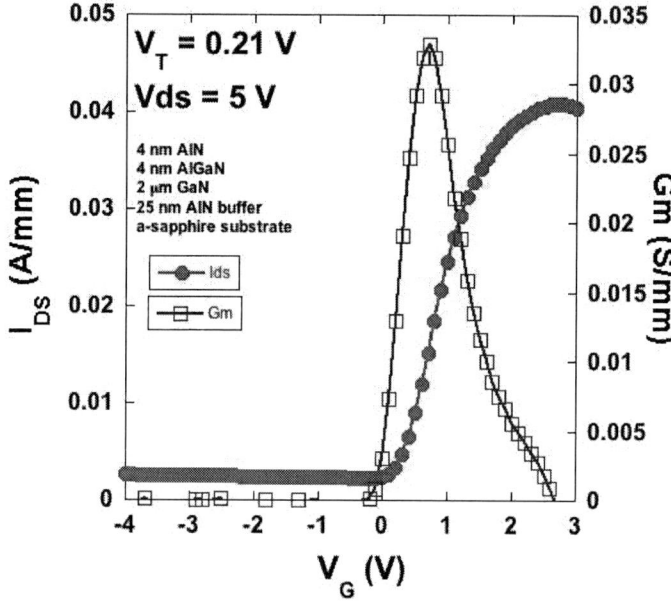

Figure 3. V_{GS}-I_{DS}-Gm curve for 4 nm AlN/4 nm AlGaN/2 μm GaN structure.

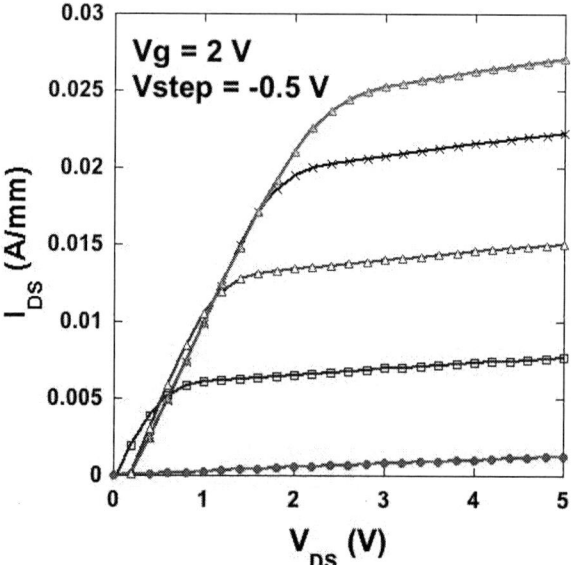

Figure 4. V_{DS}-I_{DS} curve for 4 nm AlN/4 nm AlGaN/2 μm GaN structure

Saito, et al. has detailed both theoretically and experimentally a linear relationship between threshold voltage and recessed AlGaN thickness using low power ICP etching with the following equations[1].

$$V_T = \phi_B + \frac{qN_{2D}}{\varepsilon}\left(t_{CR,G} - d\right) \qquad [1]$$

$$t_{CR,G} = \frac{\left(\Phi_B - \Delta E_C\right)\varepsilon}{qN_{2D}} \qquad [2]$$

In these equations, Φ_B represents the Schottky Barrier Height, N_{2D} represents the 2DEG density, d represents the recessed AlGaN thickness, and ΔE_C represents the conduction band offset. Using an interpolated value for ΔE_C from the literature[9], the actual 2DEG density from Hall measurements, and the barrier height extracted from source-gate I-V characteristics, the threshold voltage is accurately predicted for both the 4 and 8 nm AlGaN cases. Furthermore, this value is consistent with reports of structures using comparable AlGaN thickness[10].

Conclusion

In conclusion, a device structure has been demonstrated incorporating an ultrathin AlGaN barrier for accurate and reliable threshold voltage control, capped with a thin AlN layer in the source-drain access region to maintain high 2DEG charge. A fabrication process for this structure has been demonstrated which implements a selective wet etch of the AlN using heated photoresist developer for the gate opening step, thus bypassing plasma etching. This process has been employed using a 8 nm and 4 nm AlGaN barrier to demonstrate HEMTs with a threshold voltage of -0.48 and +0.21V, respectively, while maintaining a high mobility (>700 cm^2/V-s) and with little loss in current density relative to an unetched device.

Acknowledgments

The authors are sincerely grateful the microwave HEMT device group at NRL (S. Binari *et al.*) for insightful discussions and equipment use, the NRL Institute for Nanoscience for equipment use and support, and G. Jernigan for XPS characterization. J. K. Hite was partially supported by the ASEE. M. J. Tadjer acknowledges the scientific input of Prof. John Melngailis at the University of Maryland, College Park.

References

1. W. Saito, Y. Takada, M. Kuraguchi, *IEEE Trans Electron Devices*. 53, 2, 1-7 (2006)
2. M. Kuraguchi, Y. Takada, T. Suzuki, M. Hirose, K. Tsuda, W. Saito, Y. Saito, I. Omura. *Phys. Stat. Sol. A* 204, No. 6, 2010-2013 (2007)
3. S. Maroldt, C. Haupt, W. Pletschen, S. Muller, R. Quay, O. Ambacher, C. Schippel, F. Schwierz. *Jap. J. Appl. Phys*. 48, 04C083 (2009)
4. T. Palacios, C.S. Suh, A. Chakraborty, S. Keller, S.P. DenBaars, U.K. Mishra. *IEEE Electron Device Letters* 27, 6, 428-430 (2006)
5. Y. Cai, Y. Zhou, K.M. Lau, K.J. Chen. *IEEE Trans. Electron Devices* 53, 9, 2207-2215 (2006)
6. A. Basu, I. Adesida. *J. Appl. Phys*. 105, 033705 (2009)
7. T.J. Anderson, M.J. Tadjer, M.A. Mastro, J.K. Hite, K.D. Hobart, C.R. Eddy, Jr, F.J. Kub. *IEEE Electron Device Letters* 30, 12, 1251 (2009)
8. S. J. Pearton, J. C. Zolper, R. J. Shul, F. Ren. *J. Appl. Phys*. 86, 1-78 (1999)
9. O. Ambacher, B. Foutz, J. Smart, J.R. Shealy, N.G. Weimann, K. Chu, M. Murphy, R. Dimitrov, A. Mitchell, M. Stuzmann. *J. Appl. Phys*. 87, No. 1, 334 (2000)
10. Y. Ohmaki, M. Tanimoto, S. Akamatsu, T. Mukai. *Jap. J. Appl. Phys*. 45, No. 44, L1168-L1170 (2006)

ZnO nanorod/p–GaN heterostructured light–emitting diodes passivated using photoelectrochemical method

Jheng–Tai Yan, and Ching–Ting Lee*

Institute of Microelectronics, Department of Electrical Engineering, National Cheng Kung University, Tainan, Taiwan, Republic of China
Tel: 886–6–2082368 Fax: 886–6–2082368
E-mail: ctlee@ee.ncku.edu.tw

To fabricate the p–GaN/i–ZnO nanorod/n–ZnO nanorod (p–i–n) nanorod–heterosturctured light–emitting diodes (LEDs), i–ZnO and n–ZnO:In nanorod arrays were grown sequentially on a p–GaN layer. In order to passivate the non–radiatvie recombination centers at the nanorod sidewall, a thin $Zn(OH)_2$ layer was directly grown on the ZnO nanorod using a photoelectrochemical method. Electroluminescence emission at 386 nm was observed from the resultant LEDs. Furthermore, the light output intensity of the passivated nanorod–heterosturctured LEDs was much larger than that of the unpassivated nanorod–heterosturctured LEDs. The result demonstrated that $Zn(OH)_2$ directly grown on the ZnO nanorod sidewall could passivate the non–radiative recombination centers effectively.

Introduction

ZnO nanostructures in the heterojunction with a p–GaN layer have the benefits of simple fabrication processes and a number of superior properties, such as similar lattice constant and energy bandgap [1]. Despite impressive advances in 1D ZnO nanostructures, the high surface/volume ratio induced a high dangling bond density and a high non–radiative recombination centers. With a view to developing applications using ZnO nanorod array/p–GaN heterostructures, many studies have examined the surface non–radiative recombination centers to better understand the related physical phenomena. Specifically, the passivation of ZnO nanorods has become an important issue with regard to providing efficient nanodevices [2]. In this study, a photoelectrochemical (PEC) oxidation method was used to directly oxidize the ZnO nanorod sidewall. The optical and electrical properties of the ZnO nanorod/p–GaN LEDs were characterized by the electroluminescence (EL) spectra and current–voltage (I–V) characteristics.

Experiments

In this study, a 2 μm–thick undoped–GaN buffer layer and a 680 nm–thick Mg–doped GaN layer were grown on 350 μm–thick sapphire substrates using MOCVD system. The grown samples were thermally annealed at 750°C for 30 minutes in N_2 ambient. The activated hole concentration and mobility were 3.7×10^{17} cm^{-3} and 11.2 cm^2/V–s, respectively. A 1 mm–wide Ni/Au(20/100 nm) strip, which enclosed an area was about 1.2 mm^2, was deposited on the perimeter of the p–GaN layer. The sample was then annealed at 500°C for 10 minutes in an air ambient to perform the p–type ohmic contact. An anodic alumina membrane (AAM) template with pore's diameter of 200 nm was put on the top of the grown sample. In the vapor cooling condensation system used here [3, 4], the ZnO powder (0.85 g) and ZnO powder/In tablet (0.85 g/0.06 g) were separately put on tungsten boats and then heated. The resulting sublimated material vapor gases were driven through the pores of the AAM template using a pumping system. The material vapor gases were then condensed and deposited on the p–GaN being cooled by liquid nitrogen. Due to AAM's ideal cylindrical nanopore shape, the i–ZnO nanorod array and n–ZnO:In (In doped ZnO) nanorod array can be deposited on the p–GaN layer. The mobility and electron concentration for i–ZnO and n–ZnO were 6.0 cm^2/V–s and 7.6×10^{15} cm^{-3}, and 3.1 cm^2/V–s and 1.7×10^{20} cm^{-3}, respectively. From the scanning electron microscopy images, the density of the nanorod array was estimated to be about 1.72×10^9 cm^{-2}.

To passivate the non–radiative recombination centers at the ZnO nanorod sidewall, a passivation layer was directly grown on the ZnO nanorod sidewall using a photoelectrochemical method in an NH$_3$ chemical solution (pH value = 8) under the illumination of a He–Cd laser (power density = 10.0 mW/cm^2 and wavelength = 325 nm). Figure 1 shows the schematic diagram of the setup. The schematic energy–band diagram for the electrolytic solution/ZnO is shown in Fig. 2. A built–in electric field was induced within the depletion region due to the work function difference between the NH$_3$ electrolytic solution and the ZnO layer. Under the illumination of a He–Cd laser, electron–hole pairs were generated on the ZnO layer. The generated electrons and holes were transported to the ZnO and the NH$_3$ electrolytic solution/ZnO interface, respectively, by the built–in electric field. Since holes were accumulated at the interface, the hydroxide ion (OH$^-$) in the NH$_3$ chemical solution would react with the zinc and hole at the nanorod sidewall via the following equation:

$$Zn+2h^+ +2OH^- \rightarrow Zn(OH)_2 \quad\quad\quad (1)$$

where h$^+$ is the hole. Therefore, a Zn(OH)$_2$ layer was directly grown on the ZnO nanorod sidewall to passivate the non–radiative recombination centers.

Experimental Results and Discussions

To measure the optoelectrical characteristics, an indium–tin oxide (ITO) coated glass was put on the top of the n–ZnO:In nanorod array and heated to 150°C. The p–i–n (p–GaN layer/100 nm–high i–ZnO nanorod array/200 nm–high n–ZnO:In nanorod array) heterostructured LED is shown in Fig. 3. The current–voltage (I–V) characteristics of the p–i–n nanorod–heterostructured LEDs were measured using an HP4156C semiconductor parameter analyzer. Figure 4 (a) and (b) showed the I–V characteristics of the unpassivated and passivated nanorod–heterostructured LEDs. The I–V curves of unpassivated and passivated nanorod–heterostructured LEDs exhibited typical rectifying behavior. Because the non–radiative recombination centers of the p–i–n nanorod–heterostructured LEDs could be passivated by the PEC method, the leakage current of the passivated nanorod–heterostructured LEDs was smaller than that of unpassivated nanorod–heterostructured LEDs. It should be noted that a soft breakdown behavior from the I–V characteristics curve of the unpassivated nanorod–heterostructured LEDs was observed. This non–ideal behavior may be attributed to the parasitic resistance, such as parallel resistance, which may have existed in the nanorod–heterostructured LEDs.

The carrier concentration of the i–ZnO nanorods was much smaller than that of the p–GaN layer and n–ZnO:In nanorods for the p–i–n nanorod–heterostructured LEDs. Therefore, the depletion region of the p–i–n heterostructure mostly resided in the i–ZnO nanorod region. The electron–hole pairs, injected from the n–ZnO:In nanorod array and the p–GaN layer, were recombined in the i–ZnO nanorod and resulted in the near–band emission (NBE) at 386 nm, as shown in Fig. 5. The EL intensity of passivated p–i–n nanorod–heterostructured LEDs was stronger than that of the unpassivated p–i–n nanorod–heterostructured LEDs at the same injection current of 0.1 mA.

Conclusions

The ZnO p–i–n nanorod–heterostructured LEDs were grown using a vapor cooling condensation system. NBE emissions were found both in the passivated and unpassivated p–i–n nanorod–heterostructured LEDs. The leakage current of the passivated nanorod–heterostructured LEDs was smaller than that of unpassivated nanorod–heterostructured LEDs because the number of non–radiative recombination centers at the ZnO nanorod sidewall were reduced by the PEC directly oxidized layer. Therefore, the PEC method is expect to be an efficient way to improve the optoelectrical characteristics of the p–i–n nanorod–heterostructured LEDs.

Acknowledgments

This work was supported by the National Science Council, Taiwan, Republic of China under Grant No. NSC 98–2119–M–006–005.

References

1. A. M. C. Ng, Y. Y. Xi, Y. F. Hsu, A. B. Djurisic, W. K. Chan, S. Gwo, H. L. Tam, K. W. Cheah, P. W. K. Fong, H. F. Lui, and C. Surya, *Nanotechnol.*, **20**, 445201 (2009).
2. C. H. Lee, J. Yoo, Y. J. Doh, and G. C. Yi, *Appl. Phys. Lett.*, **94**, 043504 (2009).
3. R. W. Chuang, R. X. Wu, L. W. Lai, and C. T. Lee, *Appl. Phys. Lett.*, **91**, 231113 (2007).
4. J. T. Yan, C. H. Chen, S. F. Yen, and C. T. Lee, *IEEE Photon. Technol. Lett.*, 22, 146 (2010).

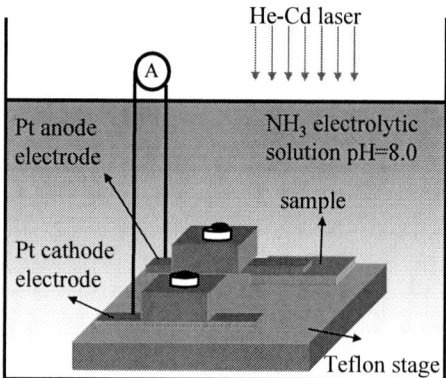

Figure 1 Schematic configuration of the photoelectrochemical oxidation system.

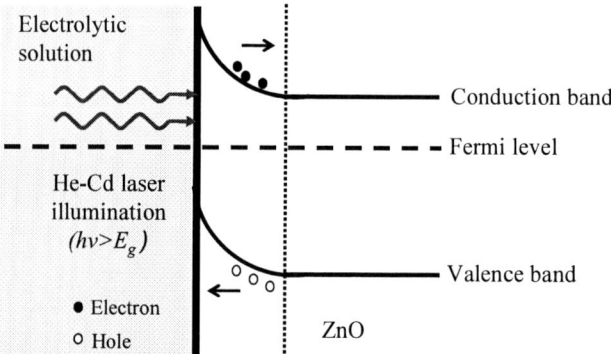

Figure 2. Schematic energy band structure for NH_3 solution/ZnO.

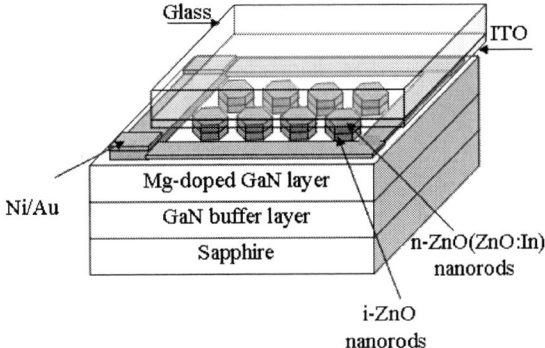

Figure 3 Schematic configuration of the p–i–n nanorod–heterostructured LEDs.

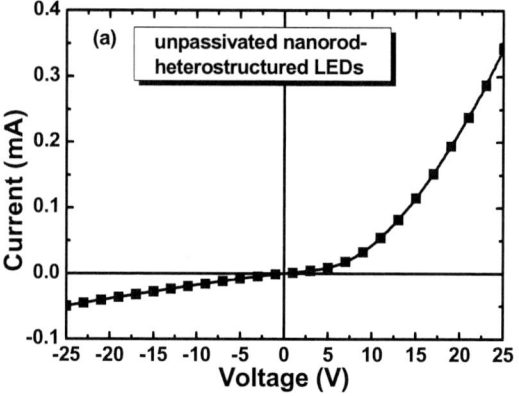

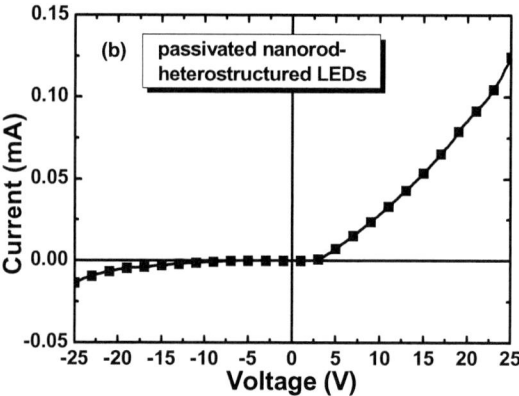

Figure 4 Room–temperature current–voltage characteristics for the (a) unpassivated and (b) passivated nanorod–heterostructured LEDs.

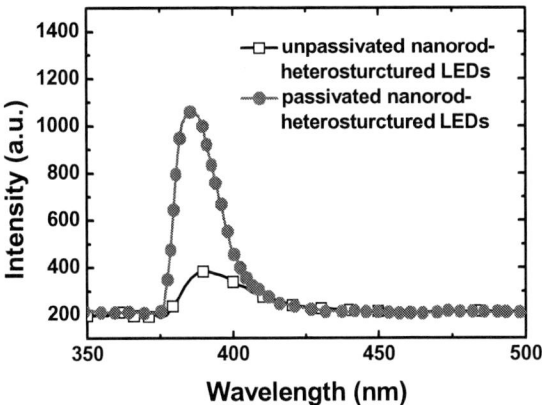

Figure 5 Room–temperature electroluminescence spectra of the unpassivated and passivated nanorod–heterostructured LEDs.

CHAPTER 3

SiC AND RELATED MATERIALS

Etch Pits of 4H-Silicon Carbide Surface Formed Using Chlorine Trifluoride Gas

Hitoshi Habuka[1], Kazuchika Furukawa[1], Keiko Tanaka[1], Yusuke Katsumi[1],
Naoto Takechi[2], Katsuya Fukae[2], and Tomohisa Kato[3]

[1] Department of Chemical and Energy Engineering, Yokohama National University,
Hodogaya, Yokohama, Kanagawa 240-8501, Japan.
[2] Shibukawa Research Laboratory, Kanto Denka Kogyo Co., Ltd.,
Shibukawa Gunma 377-8513, Japan.
[3] Energy Semiconductor Electronics Research Laboratory,
National Institutes of Advanced Science and Technology, Tsukuba, Ibaraki 305-8568,
Japan.

Surface morphology of a single-crystalline 4H-silicon carbide
(SiC) etched by chlorine trifluoride gas was studied over the wide
temperature range of 570-1570 K at atmospheric pressure in a
horizontal cold wall reactor. The etch rate of both the Si-face and
C-face 4H-SiC at the substrate temperatures between 720 and 1570
K was simultaneously measured to be nearly flat at ca. 5 μm min^{-1}.
The Si-face and C-face showed pitted surfaces at low temperatures;
the pits tended to become small and shallow with the increasing
substrate temperature. This temperature-dependent behavior is
expressed using the rate theory accounting for the slightly low
activation energy at the spot causing the pit. The etch pits formed
at low temperature may have relationship with crystalline defects,
such as threading edge dislocation and screw dislocation.

Introduction

Semiconductor silicon carbide (SiC) is an expected new material for fabricating
future power electronic devices (1). However, because of the chemical and mechanical
stability of SiC, its surface preparation is often very difficult, particularly, for obtaining a
flat surface and for removing the damaged layer caused by various mechanical processes.
Additionally, an easy technique for the etch pit formation to characterize the crystalline
defects is very useful.

For this purpose, the authors have developed a chemical etching method of SiC using
chlorine trifluoride gas (2-7). Our previous Study (7) showed that the etch pits formed at
low temperatures might have a relationship with dislocations. However, its detail should
be studied further.

In this study about a single-crystalline 4H-SiC substrate, the relationship between the
etch pits formed using chlorine trifluoride gas at low temperatures and the crystalline
defects detected by X-ray topography is discussed, along with evaluating the etch rate
and the surface morphology obtained over the wide temperature range of 570-1570 K,

Experimental

The substrate mainly used in this study was the n-type single-crystalline 4H-SiC having an on-axis (0001) surface. Nitrogen is doped at the concentration of 3 - 5 x 10^{18} cm^{-3} in this substrate. The substrate surface was prepared using a chemical mechanical polishing (8).

The horizontal cold-wall reactor shown in Fig. 1 was used in this study. This reactor consisted of a gas supply system, a quartz chamber and infrared lamps. The gas supply system has the function of providing the chlorine trifluoride gas and nitrogen gas. The height and the width of the quartz chamber are 10 mm and 40 mm, respectively.

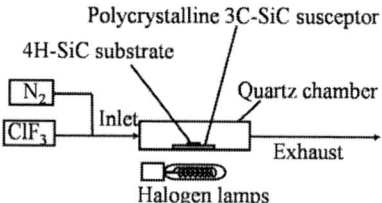

Figure 1 Horizontal cold-wall reactor for etching 4H-SiC used in this study.

Polycrystalline 3C-SiC susceptor produced by the chemical vapor deposition method was horizontally held on the bottom wall of the quartz chamber. Its dimension was 30 mm wide x 40 mm long x 0.2 mm thick. The 4H-SiC substrate, having 5 mm wide x 5 mm long x 0.4 mm thick dimensions, was placed at the center of the 3C-SiC susceptor. The 3C-SiC susceptor and 4H-SiC substrate were heated from the outside of the quartz chamber using the halogen lamps.

The entire process used in this study mainly consisted of the following three steps: (a) heating the substrate to 570-1570 K in ambient nitrogen at atmospheric pressure and at 1 slm, (b) etching the substrate surface using the chlorine trifluoride gas (>99.9%, Kanto Denka Kogyo Co., Ltd., Tokyo) at 100 % and at mainly 100 sccm, and (c) cooling the substrate to room temperature in ambient nitrogen.

The SiC etching is assumed to follow the overall reaction shown, as follows: (2),

$$3SiC + 8ClF_3 \quad 3SiF_4 + 3CF_4 + 4Cl_2. \quad\quad [1]$$

The average etch rate of the 4H-SiC by the chlorine trifluoride gas was determined by the decrease in the substrate weight. Because the 3C-SiC susceptor and the 4H-SiC substrate were simultaneously etched in the reactor, the etch rate obtained for the 4H-SiC in this study is assumed to be comparable to the average value for the 4H-SiC substrate as wide as the 3C-SiC susceptor.

The surface morphology was evaluated using an optical microscope. The X-ray topograph of Si-face and C-face 4H-SiC was taken at the beam-line BL15C of the Photon Factory of the High Energy Accelerator Research Organization (Proposal No. 2006G286), in order to evaluate the crystalline defects. .

Results and Discussion

Etch rate

Fig. 2 shows the etch rate of the Si-face and C-face 4H-SiC at the substrate temperatures between 570 K and 1570 K. The etch rate of the C-face 4H-SiC was slightly higher than that of the Si-face 4H-SiC. The etch rate of the Si-face and C-face 4H-SiC was near 5 μm min^{-1} and it was still flat at the temperatures between 770 K and 1570 K. This flat etch rate behavior is consistent with that predicted by the transport phenomena (4).

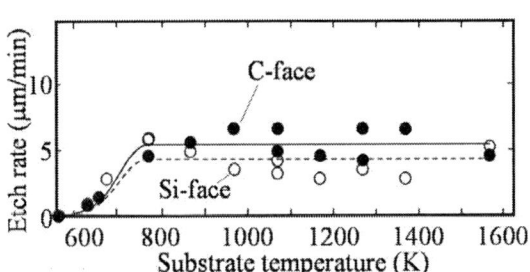

Figure 2 Etch rate of 4H-SiC using chlorine trifluoride gas at 100 %, 100 sccm and various temperatures. Dark circle: C-face, and white circle: Si-face.

Surface morphology

Si-face. Fig. 3 shows the surface morphology of the Si-face 4H-SiC before and after the etching at the chlorine trifluoride gas concentration of 100 % at various substrate temperatures and at the flow rate of 100 sccm. The etched depth was 5-18 μm.

Fig. 3 (a) shows that the surface had many small pits after the etching at 570 K. At 620 K, the pits became very large, nearly 50 μm in diameter, as shown in Fig. 3 (b). With the increasing temperature, the pits tended to become small and shallow, as shown in Figs. 3 (c) and (d). This trend became very clear at the temperatures higher than 1270 K. Fig. 3 (e) shows that the pit density significantly decreased at 1370 K.

100μ m

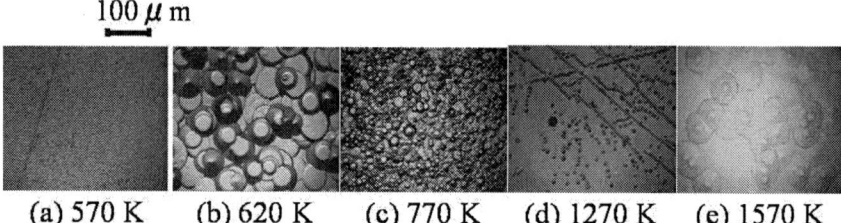

(a) 570 K   (b) 620 K   (c) 770 K   (d) 1270 K   (e) 1570 K

Figure 3 Surface morphology of Si-face 4H-SiC after the etching using chlorine trifluoride gas at the concentration of 100 %, at the substrate temperature of (a) 570, (b) 620, (c) 770, (d) 1270 and (e) 1570 K and at the flow rate of 100 sccm. This photograph was taken using an optical microscope. The etched depth was 5-18 μm.

C-face. Fig. 4 shows the surface morphology of the C-face 4H-SiC before and after the etching at the chlorine trifluoride gas concentration of 100 %, at various substrate temperatures and at the flow rate of 100 sccm. The etched depth was 10-30 μm.

100μ m

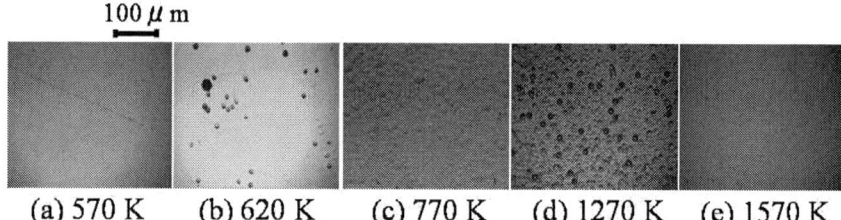

(a) 570 K   (b) 620 K   (c) 770 K   (d) 1270 K   (e) 1570 K

Figure 4 Surface morphology of C-face 4H-SiC after the etching using chlorine trifluoride gas at the concentration of 100 %, at the substrate temperature of (a) 570, (b) 620, (c) 770, (d) 1270 and (e) 1570 K and at the flow rate of 100 sccm. This photograph was taken using an optical microscope. Etched depth was 10-30 μm.

The flat surface was observed after the etching at 570 K, as shown in Fig. 4 (a), because of significantly small etch rate. However, Fig. 4 (b) shows that pits appeared at 620 K. The surface etched at the temperatures between 770 K and 1270 K had many small pits, as shown in Figs. 4 (c) and (d). The pit diameter decreased at the temperatures higher than 1370 K. The surface etched at 1570 K showed a flat surface as shown in Fig. 4 (e).

Rate theory for etch pit depth

In order to design the process producing the pit formation behavior obtained in this study, the etch pit depth is discussed following the rate theory, assuming that the etch pit is formed due to the difference of the etch rate between the perfect crystal region and the weak spot having any kinds of damage and crystalline defect (9). Following this concept, and assuming that the etchant gas concentration is the same in the perfect region and at the weak spot, the pit depth is expressed (7), as follows:

$$Pit\ depth = V_E \left(\frac{k_W - k_P}{k_P} \right) = V_E \left(exp \left(\frac{\Delta E}{RT} \right) - 1 \right) \cong V_E \frac{\Delta E}{RT} \ , \qquad [2]$$

where k_P is the rate constant of the etching in the perfect crystal region, k_W is the rate constant at the weak spot, ΔE is the difference of the activation energy between the perfect region and the weak spot, R is the gas constant, T is the substrate temperature, and V_E is the etch rate in the perfect region.

Here, assuming that the etch rate shown in Fig. 2 is the V_E in the perfect region, the normalized pit depth, h, is evaluated and shown in Fig. 5. The h value is defined using the maximum value of the pit depth, as follows:

$$h = \frac{Pit\ depth}{Pit\ depth_{MAX}} = \frac{\dfrac{V_E}{T}}{\left(\dfrac{V_{E,}}{T} \right)_{Pit\ depth_{MAX}}} \ , \qquad [5]$$

In Fig. 5, the h value at the temperatures lower than 500 K is very small; it significantly increases near 700 K. After showing its maximum, the h value gradually decreases with the increasing substrate temperature. Near 1600 K, the h value is significantly smaller than the maximum value. This trend qualitatively agrees with that of the 4H-SiC surface etched using chlorine trifluoride gas. Thus, the surface morphology trend over wide temperature range obtained in this study can be understood mainly by the rate process.

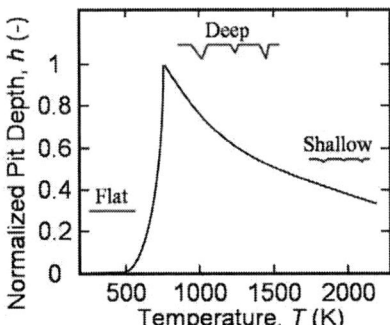

Figure 5 Normalized pit depth behavior following the rate theory.

Crystalline defect

Following the prediction from Fig. 5, the Si-face and C-face 4H-SiC surface were etched using chlorine trifluoride gas at 100 % and at 700 K. Additionally, the etch pits were compared with the X-ray topograph in order to evaluate an origin of etch pit (10, 11).

Fig. 6 (a) and Fig. 7 (a) are the X-ray topograph of the Si-face and C-face 4H-SiC surface, respectively. Fig. 6 (b) and Fig. 7 (b) are the photograph of Si-face and C-face 4H-SiC surface, respectively, etched using the chlorine trifluoride gas at 100 % and at 700 K for 60 min.

In Figs. 6 (b) and 7 (b), there were many etch pits, the positions of which corresponded to those of the spots in the X-ray topograph. The dimension of spot in Fig. 7 (a) was larger than that in Fig. 6 (a); the diameter of etch pits in Fig. 7 (b) was about 250 μm which was similarly larger than that in Fig. 6 (b), about 40 μm. Taking into account the study (12) about the dimension of etch pits formed by KOH, Figs. 6 (a) and (b) are considered to be the threading edge dislocation; Figs. 7 (a) and (b) are can be the screw dislocation.

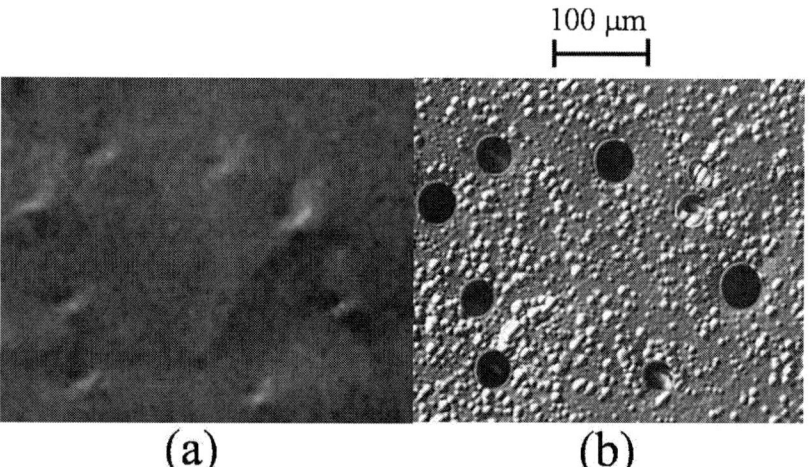

(a) (b)

Figure 6 Comparison between the X-ray topograph and the etched Si-face 4H-SiC surface. (a) X-ray topograph of the Si-face 4H SiC surface, and (b) the Si-face 4H-SiC surface etched using chlorine trifluoride gas at 100 % and at 700 K for 60 min.

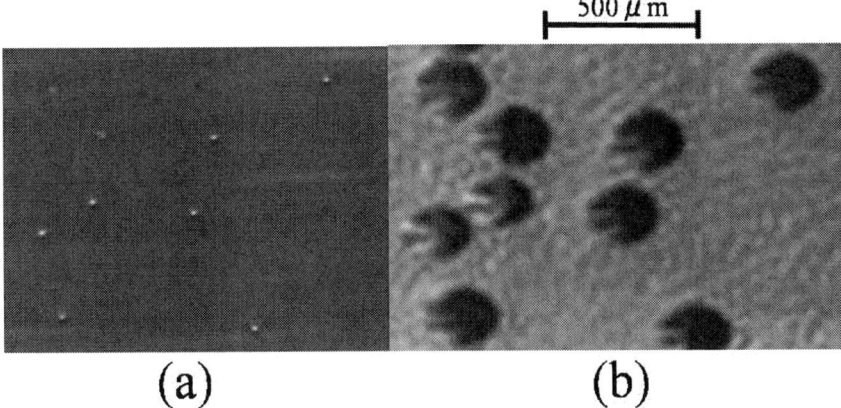

(a)   (b)

Figure 7 Comparison between the X-ray topograph and the etched C-face 4H-SiC surface. (a) X-ray topograph of the Si-face 4H SiC surface, and (b) the C-face 4H-SiC surface etched using chlorine trifluoride gas at 100 % and at 700 K for 60 min.

Conclusions

The etch rate and surface morphology of single-crystalline 4H-SiC were studied using chlorine trifluoride gas over the wide temperature range of 570-1570K at atmospheric pressure in a cold wall horizontal reactor. The etch rate of both the Si-face and C-face 4H-SiC is *ca.* 5 μm min^{-1}, and is nearly flat at the substrate temperatures higher than 720 K. The Si-face and C-face 4H-SiC showed pitted surfaces at the lower temperatures; the pits became smaller and shallower with the increasing substrate temperature. This temperature-dependent behavior of pit depth is expressed using the rate theory accounting for the slightly low activation energy at the spot causing the pit. The etch pits formed at low temperature may have relationship with the threading edge dislocation and the screw dislocation.

Acknowledgements

X-ray topography experiment has been performed under the approval of the Photon Factory Program Advisory Committee (Proposal No. 2006G286). The authors would like to thank Prof. Kenji Aramaki, Prof. Minoru Takeda, and Prof. Masahiko Aihara of Yokohama National University for their help.

References

1. M. Cooke, III-Vs Review, 18, 40 (Dec. 2005).
2. H. Habuka, S. Oda, Y. Fukai, K. Fukae, T. Takeuchi and M. Aihara, Jpn. J. Appl. Phys., 44, 1376 (2005).
3. H. Habuka., S. Oda, Y. Fukai, K. Fukae, T. Takeuchi and M. Aihara, Thin Solid Films, 514, 193 (2006).
4. Y. Miura, Y. Katsumi, S. Oda, H. Habuka, Y. Fukai, K. Fukae, T. Kato, H. Okumura and K. Arai, Jpn. J. Appl. Phys., 46, 7875 (2007).
5. H. Habuka, Y. Katsumi, Y. Miura, K. Tanaka, Y. Fukai, T. Fukae, Y. Gao, T. Kato, H. Okumura and K. Arai, Materials Science Forum, 600-603, 655 (2008).
6. Y. Miura, Y. Katsumi, K. Tanaka, S. Oda, H. Habuka, Y. Gao, Y. Fukai, K. Fukae, T. Kato, H. Okumura and K. Arai, ECS Transactions, 13 (3), 48 (2008).
7. H. Habuka, K. Tanaka, Y. Katsumi, N. Takechi, K. Fukae, and T. Kato, J. Electrochem. Soc., 156, H971 (2009).
8. T. Kato, K. Wada, E. Hozomi, H. Taniguchi, T. Miura, S. Nishizawa, K. Arai, Mat. Sci. Forum, 556-557, 753 (2007).
9. F. Shimura, Semiconductor Silicon Crystal Technology, p.244, Academic Press (San Diego, USA, 1989).
10. T. Ohno, H. Yamaguchi, S. Kuroda, K. Kojima, T. Suzuki, and K. Arai, J. Cryst. Growth, 260, 209–216 (2004).
11. X. Ma, M. Dudley, W. Vetter and T. Sudarshan, Jpn. J. Appl. Phys., 42, L1077 (2003).
12. J. Takanashi, M. Kanaya and Y. Fujiwara, J. Cryst. Growth, 135, 61 (1994),

a-plane GaN for Hydrogen Sensing Applications

Kwang Hyeon Baik[a], Wantae Lim[b], S.J. Pearton[b], Yu-Lin Wang[c], F. Ren[c], Jeongyeol Yang[d], Soohwan Jang[d]

[a] Optoelectronics Labs, Korea Electronics Technology Institute, Sungnam, Gyeonggi 463-816, Korea
[b] Department of Materials Science and Engineering, University of Florida, Gainesville, FL 32611
[c] Department of Chemical Engineering, University of Florida, Gainesville, FL 32611
[d] Department of Chemical Engineering, Dankook University, Yongin, 448-701, Korea

The performance of Pd/GaN Schottky diodes fabricated on a-plane GaN for hydrogen sensing was investigated. Pd Schottky diode on non-polar a-plane (11-20) GaN layers shows large increases in both forward and reverse bias current upon exposure to 4% H_2 in N_2. The barrier height reduction due to hydrogen exposure is 0.11 eV with long recovery times (>25 mins) at room temperature. The sensitivity to hydrogen is significantly greater than for diodes on conventional c-plane (Ga-polar) GaN.

Introduction

There is great interest in the use of nonpolar GaN due to its promise for eliminating internal electric fields, which are present in conventional c-plane III-nitrides (1). In the polar crystal orientations, both spontaneous and piezoelectric polarization electric fields cause the quantum-confined Stark Effect (QCSE), resulting in the reduction of radiative recombination rates in quantum wells (QWs) used as active layers in LEDs (2). There have been recent reports of growth optimization of non-polar GaN and of nonpolar and semipolar GaN LEDs and transistors on SiC, $LiAlO_2$, and bulk GaN substrates (3-21). There are few reports on nonpolar nitride LEDs or transistors directly grown on sapphire substrate because of the difficulty in obtaining high quality nonpolar or semipolar GaN epitaxial films (15). The growth of non-polar GaN still needs optimization, with a-plane (11-20) GaN layers grown on r-plane (1-102) sapphire containing a high density of threading dislocations (TDs) and basal stacking faults (SFs) (1, 15-18). The absence of polarization-induced surface charges should lead to differences in barrier height between the different orientations, but it is not clear how much the contact properties are influenced by defects or factors such as hydrogen adsorption (22).

The properties of Pd Schottky contacts on n-type a-plane (11-20) GaN epitaxial layers directly grown on r-plane (1-102) sapphire were investigated. Barrier heights of 0.76-0.78 eV were obtained, similar to contacts on conventional c-plane GaN. We also report the performance of Pd/GaN Schottky diodes fabricated on a-plane GaN for hydrogen sensing. Pd Schottky diodes on non-polar a-plane (11-20) GaN layers show large increases in both forward and reverse bias current upon exposure to 4% H_2 in N_2. The sensitivity to hydrogen is significantly greater than for diodes on conventional c-plane (Ga-polar) GaN and the diode characteristics remain rectifying after exposure to hydrogen.

Experimental and Discussion

For the first layer of a-plane GaN templates, 200nm thick nucleation layers were grown in N2 atmosphere at 1050°C on r-plane (1-102) sapphire wafer with -0.4~+0.4° off-axis orientations. This was followed by 700nm GaN layer with higher vertical growth rate of 0.6nm/sec and V/III ratio of 1,500 was grown at 1000°C, followed by the third layer grown for dislocation terminations. By inserting SiNx interlayer, extended defects were further reduced. A thin SiNx interlayer was used because excessive amount of SiNx makes poor surface morphology. Subsequent GaN layer was grown with higher lateral growth rate, and the ratio of lateral growth rate and vertical growth rate was maintained. The electron concentration obtained from Hall measurements was ~ 5×10^{17} cm^{-3}. For Schottky contact studies we also grew conventional c-plane GaN layers with similar thickness and doping concentration.

Ohmic contacts consisted of Ti (200Å)/Al (400Å)/Ni (200Å)/Au (800Å) deposited by e-beam evaporation. These contacts were subsequently annealed at 800°C for 60 secs under flowing N$_2$ in a Solaris 150 Rapid Thermal Processing system from Surface Science Integration. The surface was encapsulated with 2000 Å of plasma enhanced chemical vapor deposited SiN$_x$ at 300° C. Windows in the SiN$_x$ were opened by dry etching and 100 Å of Pd deposited by e-beam evaporation for Schottky contacts after UV-ozone cleaning in the same chamber and heating to remove surface oxides and carbon contamination. The samples were not exposed to air between the metal deposition and ozone cleaning process. The final metal was e-beam deposited Ti/Au (200 Å/1200 Å) interconnection contacts. Current-voltage (I-V) characteristics of the Schottky diodes were measured at 25°C in a gas test chamber in ambients of N$_2$ or 4% hydrogen in nitrogen using an Agilent 4156C parameter analyzer.

Schottky Contacts

Figure 1 shows the I-V characteristics obtained from the as-deposited Pd/Au diode in both linear and log-linear form. We fit the forward I-V characteristics to the relation for the thermionic emission over a barrier (23)

$$J_F = A^* . T^2 \exp(-\frac{e\phi_b}{kT}) \exp(\frac{eV}{nkT})$$
[1]

where J is the current density, A^* is the Richardson's constant for n-GaN, T the absolute temperature, e the electronic charge, ϕ_b the barrier height, k Boltzmann's constant , n the ideality factor and V the applied voltage. The extracted barrier heights, ϕ_b were obtained as 0.78±0.04 eV. The ideality factors were around 1.4. Note that these are identical within experimental error to the values obtained for the same contacts on c-plane GaN. These results are in contrast to the data of Kim et al. for Pd Schottky diodes on a-plane and c-plane GaN, where they found lower barrier heights on a-plane material (24). In their case, the material was doped to much higher carrier concentrations (4-6 $\times 10^{18}$ cm^{-3}) and the current-voltage characteristics were best fitted using thermionic field emission. Their ideality factors were also higher, being ~2 in all cases and they suggested that their films had a high concentration of donor –like defects on the surface.

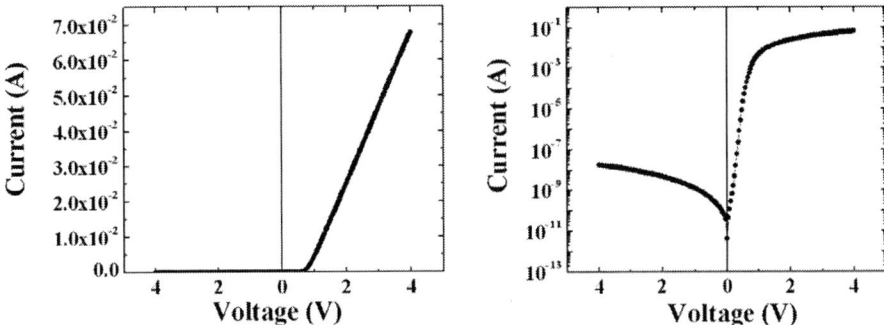

Figure 1. I-V characteristics in linear (left) or log-linear (right) form from Pd Schottky diodes on a-plane GaN

Hydrogen Sensing

Figure 2 shows I-V characteristics from the Pd/GaN a-plane diode before and after exposure to 4% H_2 in N_2. The device shows a large change in both forward and reverse currents upon exposure to the hydrogen-containing ambient. This increase in currents results from a decrease in the Schottky barrier height of the diodes. Figure 3 shows a comparison of the response to hydrogen of a-plane Pd/GaN diode with similar Pd diode fabricated on c-plane GaN. In each case, the forward bias was slightly different in order to capture the maximum response of each type of diode. The a-plane diode shows much larger response to hydrogen than conventional c-plane Ga-polar diode, but does not recover to their initial current during the 30 min period after switching back to pure N_2 ambient.

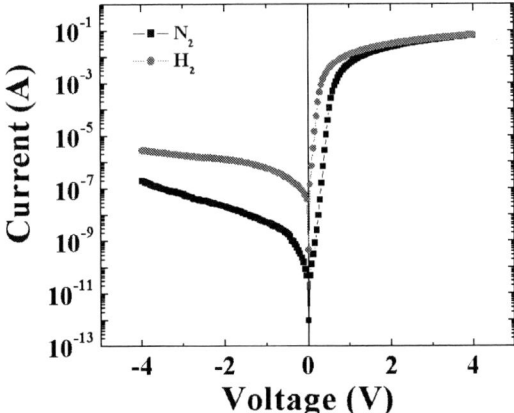

Figure 2. I-V characteristics from the Pd/GaN a-plane diodes before and after exposure to 4%H_2 in N_2

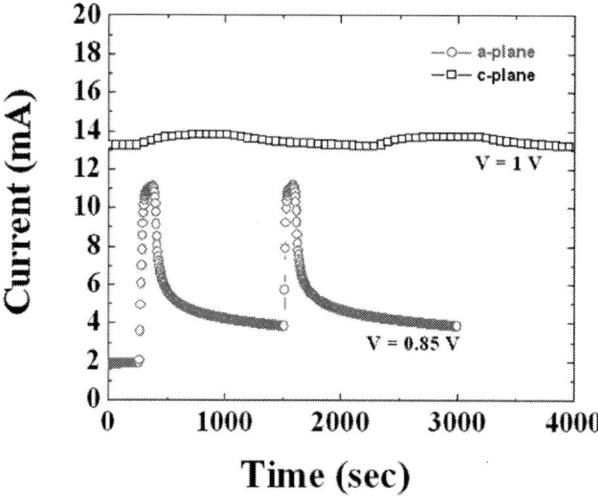

Figure 3. Comparison of the time dependent response of a-plane Pd/GaN diode with Pd diode on c-plane

Little is known about the relative stability of hydrogen bonding on a-plane GaN surfaces. Previous density functional theory for c-axis GaN suggests that hydrogen has a much higher affinity for the N-face surface of GaN than the Ga-face (25) and thus, one might expect both the orientation and the polarity of the surface to play a strong role in the hydrogen sensing characteristics. The strong affinity of the N-face of GaN for hydrogen also leads to a slower recovery of these diodes when hydrogen is removed from the ambient and it appears that for the Ga-face, the a-plane binds hydrogen more strongly than the c-plane.

Conclusion

Pd Schottky diodes on lightly doped a-plane GaN layers show similar barrier heights to those fabricated on more conventional c-axis GaN. A comparison of the hydrogen sensing characteristics of GaN Schottky diodes shows that the crystal orientation plays important role in the resulting detection sensitivity. Non-polar, a-plane diodes exhibit larger changes in barrier height than their c-plane counterparts and also slower recovery characteristics. The diode I-V characteristics remain rectifying after exposure to hydrogen.

Acknowledgment

This work was supported by 2009 Dankook university project for funding RICT.

References

1. see, for example, the special issue of MRS Bulletin on Non-Polar and Semipolar Group III Nitride-Based Materials, Vol.34, May 2009
2. C. Chen, V. Adivararahan, J. Yang, M. Shatalov, E. Kuokstis and M. Asif Khan, Jpn. J. Appl. Phys.42,L1039(2003).
3. P. Waltereit, O. Brandt, A. Trampert, H.T. Grahn, J. Meinnger, M. Ramsteiner, M. Reiche, K.H. Ploog, Nature 406, L190 (2000)
4. A. Chakaraborty, B.A. Haskell, S. Keller, J.S. Speck, S.P. DenBarrs, S. Nakamura, U.K. Mishra, Jpn. J. Appl. Phys. 44, L173 (2005)
5. K. Okamoto, H. Ohta, D. Nakagawa, M. Sonobe, J. Ichihara, and H. Takasu, Jpn. J. App. Phys. Lett. 86, L1197 (2006)
6. M. Futano, M. Ueda, Y. Kawakami, Y. Narukawa, T. Kosugi, M. Takahashi, and T. Mukai, Jpn. J. Appl. Phys. 45, L659 (2006)
7. A. Tyagi, H. Zhong, N.N. Fellows, M. Iza, J.S. Speck, S.P. DenBaars, and S. Nakamura, Jpn. J. Appl. Phys. 46, L129 (2007).
8. M. Kuroda, H. Ishida, T. Ueda and T. Tanaka, J. Appl. Phys. 102,093703 (2007).
9. Q. Sun, T.-S. Ko, C. D. Yerino, Y. Zhang, I.-H. Lee, J. Han, T.-C. Lu, H.-C. Kuo, and S.-C. Wang, Jpn. J. Appl. Phys. 48, 071002 (2009).
10. Q. Sun, Y. S. Cho, I.-H. Lee, J. Han, B. H. Kong, and H. K. Cho, Appl. Phys. Lett., 93, 131912 (2008).
11. Q. Sun, Y. S. Cho, B. H. Kong, H. K. Cho, T. S. Ko, C. D. Yerino, I.-H. Lee, and J. Han, J. Cryst. Growth 311, 2948 (2009).
12. K.-C. Kim, M.C. Schmidt, H. Sato, F. Wu, N. Fellows, M. Saito, K. Fujito, J.S. Speck, S. Nakamura, and S.P. DenBaars, Phys. Stat. Sol. (RRL) 1, No. 3, 125 (2007)
13. A. Chitnis, C. Chen, V. Adivarahan, M. Shatalov, E. Kuokstic, V. Mandavilli, J. Yang, M.A. Khan, Appl. Phys. Lett. 84, 3663 (2004)
14. X. Ni, Y. Fu, Y.T. Moon, N. Biyikli and H. Morkoc, J. Cryst. Growth, 290, 166 (2006)
15. S.-M. Hwang, Y. G. Seo, K. H. Baik, I.-S. Cho, J. H. Baek, S. Jung, T. G. Kim, and M. Cho , Appl. Phys. Lett. 95, 071101 (2009)
16. R. Miyagawa, M. Narukawa, B. Ma, H. Miyake, K. Hiramatsu, J. Cryst. Growth, 310, Issue 23, 4979 (2008)
17. S.-N. Lee, H.S. Park, J.K. Son, T. Sakong, O.H. Nam and Y. Park, J.Cryst.Growth, 307, 358 (2007)
18. T. Guhne, P. DeMierry, M. Nemoz, E. Beraudo, S. Chenot and G. Nataf, IEEE Electron. Lett. 44, No. 3 (2008)
19. B. Imer, F. Wu, M.D. Craven, J.S. Speck, S.P. DenBaars and S. Nakamura, Jpn. Appl. Phys. 45, 8644 (2006)
20. B. Bastek, F. Bertram, J. Christen, T. Wernicke, M. Weyers, and M. Kneissl, Appl. Phys. Lett. 92, 212111 (2008)
21. J-J. Huang, T-Y. Tang, C-F. Huang and C.C. Yang, J. Cryst. Growth, 310, 2712 (2008).
22. Y.-L. Wang, F. Ren, U. Zhang, Q. Sun, C. D. Yerino, T. S. Ko, Y. S. Cho, I. H. Lee, J. Han, and S. J. Pearton, Appl. Phys. Lett. 94, 212108 (2009).
23. D. K. Schroder, Semiconductor Material and Device Characterization, (Wiley and Sons, NY 1990).
24. H. Kim, S.-N. Lee, Y.Park, J.S. Kwak and T.-Y.Seong, Appl. Phys. Lett.93, 0322105 (2008).
25. J.E. Northrup and J.Neugebauer, Appl. Phys. Lett.85,3429 (2004).

Passivation of Deep Levels at the SiO$_2$/SiC Interface

A. F. Basile[a], J. Rozen[b], X. D. Chen[a], S. Dhar[c], J. R. Williams[d], L. C. Feldman[b,e], and P. M. Mooney[a]

[a] Department of Physics, Simon Fraser University, Burnaby, BC V5A 1S6, Canada
[b] Department of Physics and Astronomy, and Institute of Nanoscale Science and Engineering, Vanderbilt University, Nashville, TN 37235, USA
[c] Power Electronics R&D, Cree Inc., Durham, NC 27703, USA
[d] Department of Physics, Auburn University, Auburn, AL 36849, USA
[e] Institute of Advanced Materials, Devices and Nanotechnology, Rutgers University, Piscataway, NJ 08854, USA

The analysis of trapping phenomena in 4H- and 6H-SiC MOS capacitors from C-V and CCDLTS measurements is presented. Three categories of defect levels are distinguished: namely, oxide traps, semiconductor bulk traps, and interface states. NO annealing results in a dramatic decrease of the density of the interface states and the oxide traps in both polytypes, but does not reduce that of the SiC bulk traps.

Introduction

Silicon carbide (SiC) metal-oxide-semiconductor (MOS) transistors have the potential for high-power, high-frequency and high-temperature applications owing to attractive SiC material properties, such as the large band-gap above 3eV, bulk electron mobility comparable to that of Si, and excellent thermal conductivity. Moreover, MOS fabrication on SiC can follow Si technology, owing to the availability of silicon dioxide (SiO$_2$) as the natural oxide on both materials. However, the electrical properties of the SiC MOS system, for example the poor channel mobility compared to bulk values, are still the subject of intense research efforts. In particular, the release of C atoms during oxidation increases the density and the variety of interface states and the larger band-gap energy compared to Si makes the SiC MOS interface sensitive to a larger range of defect energy levels. The latter effect is apparent from the strong decrease in the electron channel conductivity at increasing band-gap energy, from 3eV for the 6H-SiC polytype to about 3.25eV for 4H-SiC. Post-oxidation treatments, such as NO annealing at 1175°C for 2 hours, increase the electron channel mobility in 4H-SiC to values comparable to that in as-oxidized 6H-SiC MOS [1,2]. However, the residual presence of excess C atoms in SiC near the interface was found to be inversely related to the channel mobility [3].

With the aim of relating these electrical characteristics to the interface properties, the electron trapping phenomena in n-type MOS capacitors on the (0001) Si-face of 4H and 6H-SiC polytypes were characterized by constant-capacitance deep level transient spectroscopy (CCDLTS). Compared to other interface characterization techniques, such as high-low capacitance-voltage and conductance-dispersion analysis [4], CCDLTS provides more detailed information on the physical properties of interface states, as it allows for the simultaneous extraction of energy depth and capture cross-section of the interface states. In these SiO$_2$/SiC capacitors three different types of traps, including interface states, bulk semiconductor traps and oxide traps, are characterized in terms of their response to the NO annealing passivation treatment.

Experimental

n-type 4H-SiC and 6H-SiC epilayers, 10-μm thick and doped with N at about $5 \times 10^{15} \text{cm}^{-3}$, were oxidized in flowing O_2 at 1150°C for 8 hours and subsequently underwent an Ar anneal at the same temperature for 30 min, yielding an oxide thickness of about 60nm on the (0001) Si-face. NO annealing was then performed at 1175°C, for different time lengths of 12min, 30min and120 min on the 4H-SiC samples. As previously shown, N is detected by secondary ion mass spectrometry (SIMS) only at the oxide/semiconductor interface with total densities of incorporated N atoms of 2.1, 3.5 and $6 \times 10^{14} \text{cm}^{-2}$, respectively [5]. A 4H-SiC sample without annealing has also been processed and characterized for reference. The 6H-SiC MOS capacitors were processed only in the conditions of no annealing and 120min, with a total N surface density in the latter, assumed to be similar to that measured in the corresponding 4H-SiC MOS device. 200nm-thick Al contacts, 500μm in diameter, were evaporated onto the oxide, whereas sputtered Au was used to ensure good ohmic contact to the substrate.

High-frequency C-V curves were recorded at temperatures (T) between 300 K and 80 K, from depletion to accumulation using the 1 MHz capacitance meter of the SULA DLTS spectrometer. The oxide capacitance (C_{OX}) value is obtained for voltage biases deep in accumulation, i.e. for large positive bias, and is about 130 pF for all the samples under study. The flat-band voltage (V_{FB}), i.e. the bias voltage for which the Fermi level at the oxide-semiconductor interface (F_P) is equal to the bulk Fermi energy (E_F), was extracted from each C-V curve, following the procedure explained in Ref. 6. The shift of V_{FB} with T, offset by the theoretical temperature dependence of E_F, provides an estimate of the density of interface states distributed within 0.2eV below the conduction band energy minimum (E_C). However, due to the lack of resolution in energy of this technique, only a total estimate of the interface states surface density (N_{IT}) will be obtained by using the expression:

$$N_{IT} = \frac{C_{OX} \cdot \left(V_{FB}^{80K} - V_{FB}^{300K} - \left(E_F^{80K} - E_F^{300K} \right) \right)}{qA} \quad , \qquad (1)$$

with A the area of the capacitor oxide contacts, and E_F decreasing from about -0.06eV to about -0.2eV with respect to E_C as T increases from 80K up to 300K.

DLTS spectra were obtained from the voltage transients that are generated to maintain a preset capacitance value by a feedback loop and are sampled at times t_1 and t_2. The detected voltage transients are a direct measurement of the charge trapped at interface states energetically located near E_C, within an interval determined by the filling pulse voltage (V_P) and the feedback capacitance (C_F) values. The latter was set to about 35pF, corresponding to a ratio C_F/C_{OX} of about 0.25, for all the samples investigated. The sampling times corresponded to an emission rate (e_0) of 465.1s^{-1} [7]. The filling pulse width was always set to 25 ms.

The estimate of the density of interface states as a function of energy (D_{IT}) can be obtained from the dependence of the CCDLTS signal intensity on F_P, as determined from V_P. In the simple case of an interface trap located at E_T, the amplitude of the DLTS signal will be proportional to the occupation probability f_T of the level, which is given by:

$$f_T = \frac{1}{1 + e^{\frac{E_T - F_P}{kT}}}, \tag{2}$$

having a maximum of the first derivative with respect to F_P, when

$$F_P = E_T. \tag{3}$$

The trap emission rate, e_n, is defined by

$$e_n = \sigma \cdot v_{th} \cdot N_C \cdot e^{\frac{E_T}{kT}}, \tag{4}$$

with E_T the trap energy with respect to E_C and with N_C equal to $3.25 \times 10^{15} \times T^{3/2} cm^{-3}$ and $1.73 \times 10^{15} \times T^{3/2} cm^{-3}$, for 4H-SiC and 6H-SiC polytypes, respectively. A peak in the CCDLTS spectrum occurs when

$$e_n = e_0, \tag{5}$$

which is the basic equation for standard DLTS analysis [6]. By combining equations (3)-(5), the capture cross section is related to the value of F_P corresponding to the maximum change in the DLTS signal amplitude

$$\sigma = e_0 \bigg/ \left(v_{th} \cdot N_C \cdot e^{\frac{F_P}{kT}} \right) \tag{6}$$

F_P values are extracted from a point-by-point inversion of the C-V characteristics, based on a model of the MOS capacitance as the series connection of C_{OX} and the semiconductor layer capacitance (C_D), and on the analytical calculation of the latter from the Poisson equation applied to uniformly doped layers [8].

Equations (3) and (6), can be also applied to the case of broad distributions of interface states, provided that the CCDLTS signal is monitored across F_P steps of the order of a few kT [9]. In this case, an average D_{IT} as function of energy (E_T) can be determined from the expression

$$D_{IT} = \frac{C_{OX} \cdot \Delta V}{qAkTln(t_2/t_1)}, \tag{7}$$

where ΔV is the CCDLTS signal amplitude and t_2/t_1 is equal to 2.5.

Finally, it is important to point out that the above analysis for CCDLTS spectra relies on the condition expressed in (2), which applies only to interface states, whereas CCDLTS spectra may contain emission signals from other types of defect levels, specifically traps in the SiC (bulk traps) or in the oxide near the interface (oxide traps). Nonetheless, it will be shown in the following that inconsistencies arising from the loss of validity for condition (2) can be easily identified and interpreted, thus providing general information on the spatial distribution of defects across the MOS structures.

Results

Fig. 1(a) and (b) show the CCDLTS spectra for the series of 4H-SiC and 6H-SiC capacitors obtained in saturated conditions (or nearly saturated for sample 4H-0), i.e. at V_P yielding the maximum DLTS signal. Samples are labeled according to the NO annealing time in min. As was previously reported [10], it can be seen in Fig. 1(a) that the CCDLTS signal for the 4H-0 capacitor is about one order of magnitude larger than for the 4H-120 device. The latter, in turn, has a maximum intensity very close to that of the spectra for the 6H-0 and 6H-120 MOS interfaces shown in Fig. 1(b). This trend is consistent with previous measurements of interface state density and channel electron mobility in MOS transistors [1,2]. The comparison between the effects of NO passivation on the spectra from the two polytypes also reveals that the 4H-SiC interface states are reduced quite uniformly over the entire temperature range, whereas the 6H-120 capacitor differs from the 6H-0 one, primarily in the high-temperature portion of the spectra, namely at T>150K. The latter suggests that a localized deep energy defect distribution is preferentially passivated by NO annealing in 6H-SiC.

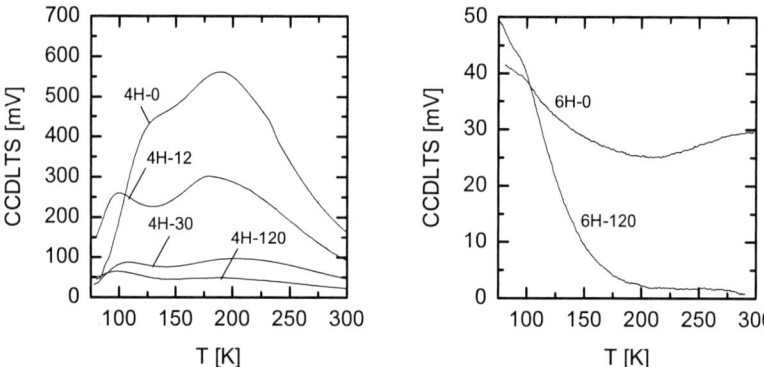

Figure 1. CCDLTS characteristics obtained at saturating V_P conditions for the (a) 4H-SiC and (b) 6H-SiC series capacitors. Samples are labeled according to the NO annealing time.

The identification of the energetic and spatial location of the defects contributing to the saturated DLTS spectra in Fig. 1 was obtained by carrying out an F_P-resolved analysis of the emission signal, as discussed in the previous section. For the sake of brevity, only data for samples 4H-30 and 6H-0 MOS will be shown as examples of this analysis.

4H-30 MOS

Fig 2(a) shows the C-V characteristics, at temperatures of 80K and 300K, along with those at two intermediate temperatures 100K and 200K. V_{FB} increases from 0.1V around 300K up to 1.2V at 80K. From expression (1), a rough estimate of the total amount of trapped charge within about 0.2eV below E_C is $N_{IT} = 4 \times 10^{11} cm^{-2}$.

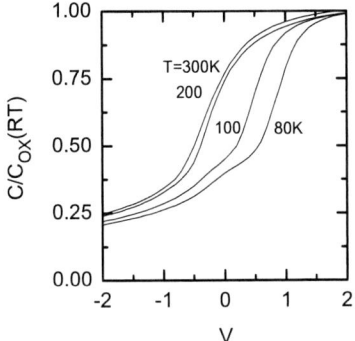 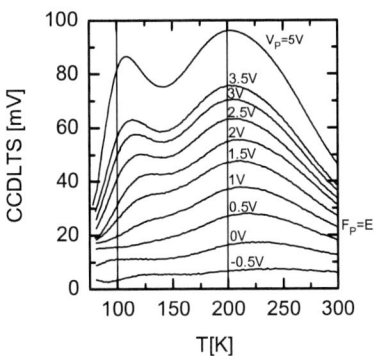

Figure 2. Experimental data for the 4H-30 capacitor. (a) High-frequency C-V characteristics at the extremes of the measurements temperature range, and at the DLTS peak temperature values. (b) CCDLTS spectra at different filling-pulse conditions. The label $F_P = E_F$ indicates the flat-band bias conditions at 300K.

Two peaks in Fig. 2(b), located at 100K and 200K, are clearly visible. The behavior of the DLTS signal with F_P at temperature 200K can be thoroughly appreciated. This peak is already apparent when the sample is biased in depletion, i.e at $V_P < 0V$ ($F_P < E_F$), and it strongly increases in amplitude as V_P is increased to accumulation ($F_P > E_F$). This indicates that the corresponding D_{IT} extends over a wide range of energies and it is peaked within 0.13eV below E_C. However, from expression (6), it was found that σ varies over several orders of magnitude from $10^{-16}cm^2$, corresponding with the signal below F_P, up to $10^{-22}cm^2$, for the portion of the spectra above E_F. While the former value may be associated with interface traps, the latter is clearly too small to represent a physical capture cross-section. This anomalous value can be understood if the condition specified by equation (2) is not valid for these traps, i.e. if they are located in the oxide, but close to the interface where trapped electrons are emitted to the 4H-SiC conduction band.

At low T, namely around 100K, a peak is generated only for $F_P > E_F$, whereas rather featureless spectra are obtained for $F_P < E_F$. The largest increase of the latter signal with increasing F_P occurs at around $V_P = 0V$ ($F_P = -0.3eV$) and it saturates around $V_P = 1V$ ($F_P = E_F = -0.1eV$), i.e. when the sample is biased in depletion. The application of expression (6) to estimate the capture cross-section yields values of σ which are orders-of-magnitude larger than the upper physical limit of $10^{-13}cm^2$. This again indicates that equation (2) cannot be applied to these traps. That trapping is detected at very low T, suggests that these traps must be energetically as shallow as E_F, The range of F_P values indicates that the traps are located in the SiC (semiconductor bulk traps) and are distributed near the MOS interface over a depth of ~100 nm, the latter corresponding to the depletion width for F_P at -0.3eV.

Estimates of the average D_{IT} for all the trap categories were obtained at the two peak temperatures. At 200K, interface states account for a D_{IT} of about $5 \times 10^{11}cm^{-2}eV^{-1}$, from the signal below flat-band, whereas oxide traps feature a density of $2.4 \times 10^{12}cm^{-2}eV^{-1}$, from the rest of the intensity above E_F. Similarly, at 100 K, a D_{IT} of ~$4 \times 10^{12}cm^{-2}eV^{-1}$ is obtained for oxide traps, from the portion of the spectrum above flat-band. Finally, an equivalent D_{IT} of $1 \times 10^{12} cm^{-2} eV^{-1}$ was obtained for the semiconductor bulk traps described above, showing that their impact on the trapping phenomena as detected by CCDLTS is nearly comparable to the dominant effect by the oxide traps.

6H-0 MOS

The comparison of the trapping phenomena in the 6H-SiC capacitor without post-oxidation annealing and the 4H-30 sample, can provide additional insight about the effectiveness of NO annealing with respect to the nature and energy location of interface traps. In fact, the similarity in transport properties repeatedly reported for transistors fabricated from as-oxidized 6H-SiC and NO-annealed 4H-SiC indicates that the energy position of the large majority of the interface defects that are passivated by NO treatment are resonant with the 6H-SiC conduction band and must therefore be less than ~0.2 eV below the 4H-SiC conduction band [11]. In agreement with this, the temperature dispersion of the C-V characteristics for the 6H-0 capacitor in Fig. 3(a) shows a V_{FB} variation from -0.8V at 300K up to 0V at 80K, and yields $N_{IT} = 2.7 \times 10^{11} cm^{-2}$, comparable to that estimated for the 4H-30 capacitor.

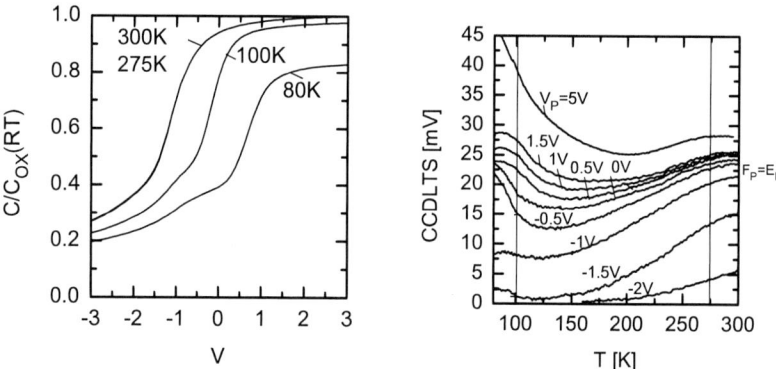

Figure 3. Experimental data for the 6H-0 capacitor. (a) High-frequency C-V characteristics at the extremes of the measurement temperature range, and at the DLTS peak temperature value. (b) CCDLTS spectra at different filling-pulse conditions. The label $F_P = E_F$ indicates the flat-band bias conditions at 300K.

CCDLTS spectra in Fig. 3(b) show that, despite the general similarities in the electrical properties, the spectroscopic composition of the total N_{IT} differs noticeably from that presented here for the 4H-30. In particular, the high-temperature portion of the spectra in Fig. 3(b), along the vertical line at T = 275K, displays a dependence on F_P that closely follows Eq. (2). The DLTS signal shows a large increase for $V_P = -1V$ ($F_P = -0.5eV$) and saturates around $V_P = -0.5V$ ($F_P = E_F = -0.2eV$). Moreover, the estimate of the capture cross-section by expression (6) from these parameters, namely trap energy depth of 0.5eV and peak temperature of 275K, yields σ values of $\sim 10^{-15}$ cm^2. Similar values have been already obtained for interface states distributed in the upper half of 6H-SiC band-gap by long-time-constant CCDLTS, that have been associated with graphite-like carbon-clusters [12].

The low temperature portion of the spectra in Fig. 3(b) shows large similarities with the corresponding part in Fig. 2(b) for 4H-30. In particular, two different dependences of the CCDLTS amplitude on F_P can be identified. Namely, the largest increase occurs at around $V_P = 0V$ ($F_P = -0.3eV$) and saturates around $V_P = 1V$ ($F_P = E_F = -0.1eV$), thus appearing as the signature of shallow bulk traps spatially extending over 100nm from the interface, as described for the 4H-30 sample. For $F_P>E_F$, the F_P dependence typical of oxide traps is observed. Contrary to the 4H-SiC capacitors, the oxide trap peak position could not be determined within the temperature range and emission rates available with the spectrometer, i.e. at T higher than 80K and e_0 lower than $465.1s^{-1}$. That we detect only the low energy tail of this trap distribution is consistent with the assumption that the majority of these trapping centers lie above E_C in 6H-SiC.

Estimates for the average D_{IT} related to both the oxide traps and the SiC traps yield values of $\sim 1.2\times 10^{12} cm^{-2} eV^{-1}$, thus showing that, as for 4H-30 MOS, trapping phenomena by semiconductor bulk traps may be as important as those by oxide defects. In addition, the deep interface states, arising from carbon clusters, would account for a D_{IT} of 4×10^{11} $cm^{-2} eV^{-1}$.

Summary and Conclusions

Table I reports for each sample the evolution of the integrated N density at the SiO_2/SiC interface [5], which is assumed to be similar for both 4H- and 6H-SiC, along with the N_{IT} extracted from the C-V measurements and D_{IT}, corresponding to the CCDLTS signals for the three categories of traps discussed in the previous section: namely, oxide traps yielding CCDLTS peaks for $F_P>E_F$ at 100K, interface states corresponding to surface band-bending at $F_P<E_F$ above 200K and semiconductor bulk traps corresponding to the DLTS signal at low T for $F_P<E_F$.

TABLE I. Evolution of the N-atoms incorporation, of N_{IT} of the D_{IT} for three different trap groups with NO annealing

Sample	N-SIMS $10^{14} cm^{-2}$	N_{IT} $10^{11} cm^{-2}$	Interface states $10^{12} cm^{-2} eV^{-1}$	Oxide Traps $10^{12} cm^{-2} eV^{-1}$	SiC Bulk Traps $10^{12} cm^{-2} eV^{-1}$
4H-0	0	16	3	20	-
4H-12	2.1	13	2	14	-
4H-30	3.5	4	0.5	1.4	1.0
4H-120	6	3.9	0.5	1.5	1.5
6H-0	0	2.7	0.4	1.2	1.2
6H-120	6	6.4	<0.1	0.3	2.4

The identification of the three different categories by analysis of the F_P-resolved DLTS, was based on the fact that only traps observed for $F_P<E_F$ at T above 200K, feature a capture cross section typical of interface states. In contrast, the other two categories appear to have anomalous capture cross section values that can be accounted for only if Eq. (2) is not valid. Therefore, those signals were interpreted as corresponding to spatial distributions of defects in the oxide or in the SiC near the interface. As for the signal detected below 100K for F_P around -0.3eV, it can be understood as likely originating from energetically shallow trap levels distributed across 100nm in the depletion region of the SiC epilayer.

N incorporation at the SiO_2/SiC interface results in a decrease of the D_{IT} of the interface states and the oxide traps by about one order of magnitude in 4H-SiC, but only by a factor of about four in 6H-SiC. In contrast, the reduction of total N_{IT}, obtained from

C-V measurements, is less than a factor of four in 4H-SiC and is negligible in 6H-SiC capacitors, suggesting that all the electron traps are not detected by this method. The magnitude of the residual D_{IT} in 4H-SiC after two hours of annealing time, when N incorporation is essentially saturated [5], closely resembles that in as-oxidized 6H-SiC devices, consistent with the general similarity between the electrical characteristics of these two types of MOS transistors. This confirms that the effect of the wider 4H-SiC bandgap can be overcome by NO annealing. However, this residual D_{IT}, is still of concern for channel transport properties of SiC MOS transistors. We note that the incorporation of N atoms does not seem to affect the concentration of the low-temperature traps that are dominant in the 6H-SiC samples and which are also observed in nitrided 4H-SiC samples. These traps are likely energetically shallow defects spread over tens of nm in the SiC adjacent to the interface and thus may be related to the presence of excess C in SiC.

Acknowledgments

This work was supported in part by the Canadian National Sciences and Engineering Council through Discovery Grant RGPIN/311687-2005, and the U.S. Army (ARL and TACOM/TARDEC) through Research Grants No. W911NF-07-2-0046 and W56HZV-06-C-0228.

References

1. G.Y. Chung, C.C. Tin, J.R. Williams, K. McDonald, R.K. Chanana, R.A. Weller, S.T. Pantelides, L.C. Feldman, O.W. Holland, M.K. Das, and J.W. Palmour, *IEEE Electron Dev. Lett.* **22**, 176 (2001)
2. Y. Deng, W. Wang, Q. Fang, M.B. Koushik, and T.P. Chow , *J. Electron. Mater.* **35**, 618 (2006)
3. T.L. Biggerstaff, C.L. Reynolds, T. Zheleva, A. Lelis, D. Habersat, S. Haney, S.-H. Ryu, A. Agarwal, and G. Duscher, *Appl. Phys. Lett.* **95**, 032108 (2009)
4. E.H. Nicollian, and J.R. Brews, *MOS Physics and Technology* (Wiley, Hoboken, NJ, 2003).
5. J. Rozen, S. Dhar, M.E. Zvanut, J.R. Williams, and L.C. Feldman, *J. Appl. Phys.* **105**, 124506 (2009)
6. S. Dhar, X.D. Chen, P.M. Mooney, J.R. Williams, and L.C. Feldman, *Appl. Phys. Lett.* **92**, 102112 (2008)
7. D.V. Lang, JAP **45**, 3023 (1974)
8. S.M. Sze, *Physics of Semiconductor Devics* (Wiley-Interscience, New York, 1969)
9. N.M. Johnson, J. Vac. Sci. Technol., **21**, 303 (1982)
10. X.D. Chen, S. Dhar, T. Isaacs-Smith, J.R. Williams, L.C. Feldman, and P.M. Mooney, *J. Appl. Phys.* **103**, 033701 (2008)
11. R. Schoerner, P. Friedrichs, D. Peters, and D. Stephani, *IEEE Electron Dev. Lett.* **20**, 241 (1999)
12. M. Bassler and G. Pensl: *Mat. Sci. Engin. B* Vol. 61-62, p. 490 (1999)

CHAPTER 4

GENERAL POSTER SESSION

ECS Transactions, 28 (4) 105-110 (2010)
10.1149/1.3377106 ©The Electrochemical Society

pnpn and *npn* Heterostructural Optoelectronic Devices

Der-Feng Guo[a]

[a] Department of Electronic Engineering, Air Force Academy, Kangshan, Kaohsiung 820, Taiwan

An ohmic-contact *pnpn* triangular-barrier optoelectronic switch (TBOS) has been fabricated. The triangular barrier is formed by inserting an InGaAs p-type delta-doped ($\delta(p^+)$) quantum well between two n^--GaAs layers. Due to the hole accumulation in the $\delta(p^+)$ well and the avalanche multiplication in the reverse-biased region, an S-shaped negative differential resistance (NDR) is obtained in the device characteristics. With a replacement of the ohmic-contact p-type cap layer with a Schottky-contact n-type layer in the *pnpn* TBOS, a Schottky-contact *npn* TBOS has also been proposed. A double S-shaped NDR phenomenon is observed in the *npn* TBOS characteristics due to the hole accumulation in the $\delta(p^+)$ well and the sequential avalanche multiplications in the reverse-biased region and Schottky-contact junction. Both devices show a flexible optical function as a result of the triangular barrier heights associated with incident light.

Introduction

Possessing potential applications in microwave oscillation and digital circuits, switches with negative-differential-resistance (NDR) characteristics have been devices of great features in the recent two decades (1-5). Optoelectronic switches have also been applied widely in the areas of optoelectronic integrated circuits (OEICs) and optical computing for their photonic and electronic switching capabilities. Several optoelectronic switches based on *pnpn* structure have been proposed (6, 7). These devices have bistable states switchable by optical or electrical input due to the hole accumulation at the potential minimum and the avalanche multiplication in the reverse-biased region. However, minority-carrier storage might slow down the response speed of these devices. Furthermore, the bistable states could not effectively reduce the number of elements and process step in actual applications. In this work, an ohmic-contact *pnpn* triangular-barrier optoelectronic switch (TBOS) is proposed. The triangular barrier, promising for minimizing the minority-carrier storage (8-11), is formed by inserting an InGaAs p-type delta-doped ($\delta(p^+)$) quantum well between two n^--GaAs layers. Due to the hole accumulation in the $\delta(p^+)$ well and the avalanche multiplication in the reverse-biased region, an S-shaped NDR is obtained in the device characteristics. In order to achieve multiple operation states for simplifying the actual applications, a Schottky-contact *npn* TBOS is also proposed. The Schottky-contact *npn* TBOS has the same structure as the ohmic-contact *pnpn* TBOS except the ohmic-contact p-type cap layer is replaced by a Schottky-contact n-type layer. A double S-shaped NDR phenomenon is observed in the Schottky-contact *npn* TBOS characteristics due to the hole accumulation in the $\delta(p^+)$ well and the sequential avalanche multiplications in the reverse-biased region and

105

Schottky-contact junction. Both devices also show a flexible optical function as a result of the triangular barrier heights associated with incident light.

Experiments

The devices were grown by MBE on (100)-oriented n^+-GaAs substrates. The ohmic-contact $pnpn$ TBOS, named device TBOS-1 hereafter, contained a 200nm n^+-GaAs (3×10^{18} cm^{-3}) buffer layer, a 600nm n^--GaAs (2×10^{16} cm^{-3}) layer, a 15nm undoped In$_{0.2}$Ga$_{0.8}$As quantum well, a 600nm n^--GaAs (2×10^{16} cm^{-3}) layer, and a 200nm p^+-GaAs (5×10^{18} cm^{-3}) cap layer. A $\delta(p^+)$ (5×10^{13} cm^{-2}) sheet was inserted in the center of the InGaAs quantum well. The structure of the Schottky-contact npn TBOS, named device TBOS-2, was similar to that of device TBOS-1 except the p^+-GaAs cap layer was replaced by an n-GaAs (3×10^{17} cm^{-3}) layer with the same thickness. After the MBE growth, metal contacts were achieved by conventional evaporation, lift-off and alloying techniques. AuZn and Au were employed as the ohmic-contact and Schottky-contact metals for the p^+-GaAs cap layer of device TBOS-1 and the n-GaAs cap layer of device TBOS-2, respectively. The ohmic and Schottky contacts were in the form of a ring, which was enlarged on one side to enable bonding. The open area for incident-light optical window had a diameter of 30μm. A circular mesa, 80μm in diameter, was made by etching down to the substrate for both devices. AuGe was employed as the ohmic metal for the substrates.

Results and Discussion

The device operations can be explained by reference to band diagrams. Figures 1(a) and (b) show the band diagrams of devices TBOS-1 and TBOS-2 with a positive V_{CE} voltage, respectively. The InGaAs $\delta(p^+)$ well and n^--GaAs barrier layer and n^--GaAs active layer achieve a triangular barrier for both devices. For device TBOS-1, the p^+-GaAs cap layer is the collector. But the n-GaAs cap layer of device TBOS-2 is the emitter. In device TBOS-1, focusing a light source of energy higher than the band-gap energy of GaAs onto the device surface will generate electron-hole pairs in the device. With a positive V_{CE} voltage applied to the device, the photogenerated electrons will travel toward the collector, while the photogenerated holes will move toward the emitter and part of the holes will be accumulated in the $\delta(p^+)$ well, as shown in Fig.1(a). The accumulated holes will increase the two-dimensional free holes concentration P_{2D} and compensate for some of the two-dimensional acceptor ions density Q_{2D} in the $\delta(p^+)$ sheet. According to that the potential barrier ϕ is proportional to $Q_{2D} - qP_{2D}$ (12), the ϕ will reduce, which will influence the switching characteristics of device TBOS-1. In the high-impedance OFF state, the emitter and collector junctions are both slightly forward biased, but the active layer is reversely biased. Due to high impedance of the depletion region of the active layer, almost all voltage is dropped in the active layer. As the applied voltage is increased to the switching voltage V_s, the strong electric field across the active layer will be sufficient to cause an avalanche multiplication. The multiplied holes will move toward the emitter, and some of the holes will also be accumulated in the InGaAs

$\delta(p^+)$ well to lower the potential barrier ϕ. This potential redistribution process will cause an S-shaped NDR phenomenon in the experimental current-voltage (I-V) characteristics. Then, a low-impedance ON state is obtained. Because the characteristics of device TBOS-1 are influenced by the input-light power, the switching is controllable by changing the input light.

For device TBOS-2 supplied with a light source and a positive V_{CE} voltage, the photogenerated electrons will travel toward the collector, while the photogenerated holes will move toward the emitter and part of the holes will be accumulated in the $\delta(p^+)$ well, as shown in Fig.1(b). The accumulated holes will also reduce the potential barrier ϕ. If the applied voltage V_{CE} is high enough, the electric field across the reverse-biases active layer will cause the first avalanche multiplication. The multiplied electrons will travel toward the collector, and the multiplied holes will move toward the emitter. Part of the holes will also be accumulated in the $\delta(p^+)$ well to lower the potential barrier ϕ, and the electrons injected from the emitter to collector will, therefore, increase. This potential redistribution process will cause the first S-shaped NDR phenomenon in the experimental I-V characteristics. If the V_{CE} voltage is increased further, the voltage drop across the reverse-biased Schottky-contact junction will increase due to the accumulated holes in the $\delta(p^+)$ well. Then the second avalanche multiplication can appear in the Schottky-contact junction. And the second S-shaped NDR phenomenon will be observed in the I-V characteristics. Therefore, device TBOS-2 will display three different operation states, i.e., the initial off-state, intermediate on-state and final on-state. Because the characteristics of device TBOS-2 are influenced by the input-light power, the switching is also controllable by changing the input-light.

Figures 2(a) and (b) show the I-V characteristics of devices TBOS-1 and TBOS-2, respectively. The solid lines are the characteristics under dark and dotted lines the characteristics under illumination. A proposed bias circuit is illustrated in the inset of Fig.2(a) and load lines also depicted in Figs. 2(a) and (b). Single and double S-shaped NDR performances are respectively observed in devices TBOS-1 and TBOS-2 characteristics under both dark and illumination conditions. For device TBOS-2 under illumination, lower values of V_{CE} voltage are sufficient to achieve the switching in the initial off-state (V_{S1}) and intermediate on-state (V_{S2}) due to the photogenerated holes accumulated in the $\delta(p^+)$ well to lower the potential barrier ϕ. Also owing to the accumulated holes lowering the barrier, the switching currents in the initial off-state (I_{S1}) and intermediate on-state (I_{S2}) are larger under illumination. The smaller holding voltages and holding currents in the intermediate on-state (V_{H1}, I_{H1}) and final on-state (V_{H2}, I_{H2}) under illumination are due to the internal gains increased by the photogenerated holes accumulated in the $\delta(p^+)$ well (13-15). The reasons for V_{S1}, I_{S1}, V_{H1}, I_{H1} variations with light mentioned above are also applied to device TBOS-1.

In addition to the switching characteristics controllable by the input-light, we can observe the optical switching by biasing the TBOSs to just below turn-on and then introducing light to turn the TBOSs on, or by biasing the TBOSs to just before turn-off and removing the light source to turn the TBOSs off. In the dark, the load lines intersect the characteristics at A only for both devices, as shown in Figs. 2(a) and (b). This means that the TBOSs are stable in the off state in the dark. In the illumination case, it is seen that the stable point is at B for device TBOS-1 and is at B and C for device TBOS-2 in the on states. The TBOSs will make a transition from A to B or C promptly when light falls on the devices. As the light intensity drops, the characteristics revert to the dark situation and the TBOSs will switch back to A. Therefore, both devices can be switched on and off with an optical input.

Conclusion

An ohmic-contact *pnpn* triangular-barrier optoelectronic switch (TBOS) was fabricated. An S-shaped NDR was obtained in the device characteristics. With a Schottky-contact *n*-type layer to replace the ohmic-contact *p*-type cap layer in the ohmic-contact *pnpn* TBOS, a Schottky-contact *npn* TBOS was also proposed to obtain a double S-shaped NDR in device characteristics. Attributed to the triangular barrier heights associated with incident light, both devices showed a flexible optical function.

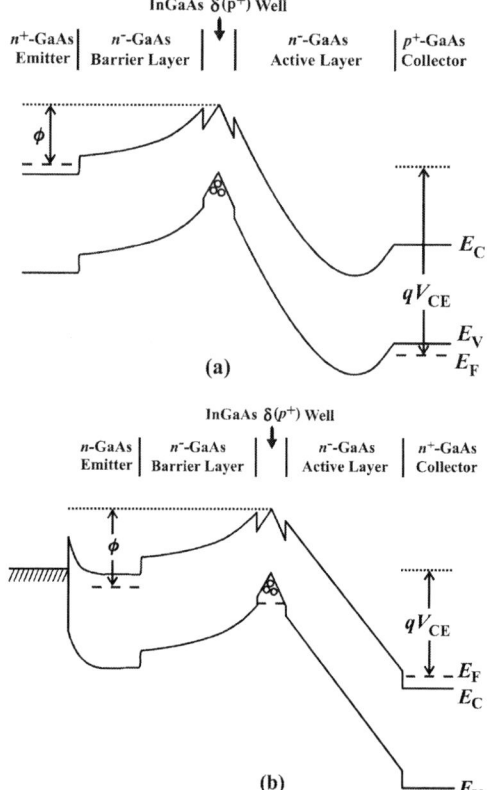

Figure 1. Band diagrams of devices (a) TBOS-1 and (b) TBOS-2 with a positive V_{CE} voltage.

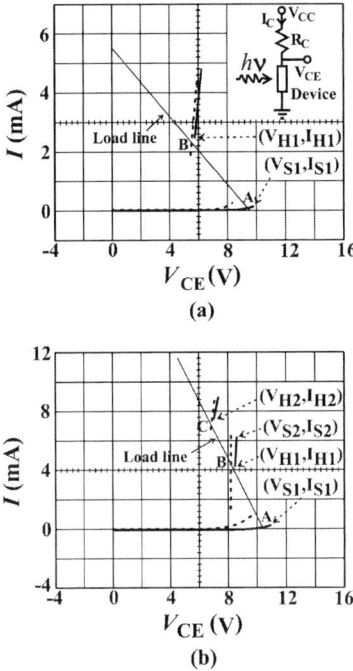

(a)

(b)

Figure 2. *I-V* characteristics of devices (a) TBOS-1 and (b) TBOS-2. The solid lines are the characteristics under dark and dotted lines are the characteristics under illumination. A proposed bias circuit is illustrated in the inset of Fig.2(a) and load lines also depicted in Figs. 2(a) and (b).

Acknowledgments

This work was supported by the National Science Council of the Republic of China under Contract No. NSC 98-2221-E-013-001.

References

1. K. J. Chen, T. Waho, K. Maezawa and M. Yamamoto, *IEEE Electron Device Lett.*, **17,** 309 (1996).
2. J. Shen, S. Tehrahi, H. Goronkin, G. Kramer and R. Tsui, *IEEE Electron Device Lett.,* **17,** 94 (1996).
3. J. Kastrup, H. T. Grahn, K. Ploog, F. Prengel, A. Wacker and E. Schöll, *Appl. Phys. Lett.,* **65,** 1808 (1994).
4. M. J. Kearney, A. Condie and J. Dale, *Electronic Lett.*, **27,** 721 (1991).
5. H. Sakata, K. Utaka and Y. Matsushima, *Electronics Lett.,* **30,** 1792 (1994).
6. F. Y. Huang and H. Morkoc, *Appl. Phys. Lett.,* **64,** 405 (1994).
7. G. W. Taylor, J. G. Simmons, A. Y. Cho and R. S. Mand, *J. Appl. Phys.,* **59,** 596 (1986).

8. A. Mccowen, S. B. H. Shaari and K. Board, *IEE Proceedings,* **135,** 107 (1988).
9. P. K. Rees and J. A. Barnard, *IEEE Trans. Electron Devices,* **32,** 1741 (1985).
10. S. Hutchinson , M. Carr, R. Gwilliam, M. J. Kelly and B. J. Sealy, *Electronic Lett.,* **31,** 583 (1995).
11. K. Board and M. Darwish, *Solid-State Electronics,* **25,** 529 (1982).
12. R. J. Malik, T. R. AuCion, R. L. Ross, K. Board, C. E. C. Wood and L. F. Eastman, *Electronic Lett.,* **16,** 836 (1980).
13. G. W. Taylor, R. S. Mand, J. G. Simmons and A. Y. Cho, *Appl. Phys. Lett.,* **49,** 1406 (1986).
14. J. G. Simmons and G. W. Taylor, *IEEE Trans. Electron Devices,* **34,** 973 (1987).
15. S. J. Kovacic, J. G. Simmons, J. P.Noel, D. C. Houghton and S. Kechang, *IEEE Electron Device Lett.,* **12,** 439 (1991).

ECS Transactions, 28 (4) 111-118 (2010)
10.1149/1.3377107 ©The Electrochemical Society

npn **Heterostructural Optoelectronic Switch with Collector Confinement Layer**

Der-Feng Guo[a]

[a] Department of Electronic Engineering, Air Force Academy, Kangshan, Kaohsiung 820, Taiwan

In order to achieve high optical sensitivity and low holding power, a wide-gap carrier confinement layer was introduced into the collector region of an *npn*-heterostructure optoelectronic switch. A similar device without the confinement layer was also fabricated to demonstrate the performance improvement. Both devices were found to have bistable electrical states: a high-impedance OFF state connected to a low-impedance ON state by a region of negative differential resistance (NDR). The functional characteristics were based on the avalanche multiplication.

Introduction

The optoelectronic switch is a device with two distinct stable states, i.e., a high-impedance OFF and a low-impedance ON states. Switching from one state to the other can be induced by either optical or electrical input (1-5). In two-dimensional arrays for parallel optical information processing, the optoelectronic switch is a primary device (6, 7). To avoid the excess heat dissipation, low holding power is required in the characteristics of the device. To achieve the high optical fanout, high light output and high optical sensitivity are also essential. Since the light output is limited by the heat dissipation, improving the optical sensitivity is optimized (8). With high gains and high-speed operations, *npn* heterostructures can be used to produce promising optoelectronic switches (9-12). In this work, *npn* heterostructures are exploited as optoelectronic switches. To improve the device performance, a wide-gap carrier confinement layer is introduced into the collector region of one of the studied devices. The device with the confinement layer presents a lower holding power and a higher optical sensitivity than the one without the confinement layer.

Experiments

The structure of the device with a confinement layer in the collector region, numbered device G220, consisted of a 300nm n^+-GaAs (3×10^{18} cm^{-3}) buffer layer, a 200nm *i*-GaAs layer, a 50nm N-Al$_{0.4}$Ga$_{0.6}$As (8×10^{17} cm^{-3}) confinement layer, a 50nm *n*-GaAs (8×10^{17} cm^{-3}) layer, an 80nm p^+-GaAs (8×10^{18} cm^{-3}) layer, a 200nm N-Al$_{0.4}$Ga$_{0.6}$As (5×10^{17} cm^{-3}) layer and a 200nm n^+-GaAs (3×10^{18} cm^{-3}) cap layer. The structure of the device without the confinement layer, numbered device G330, was the same as that of device G220 except that the 50nm N-Al$_{0.4}$Ga$_{0.6}$As (8×10^{17} cm^{-3}) confinement layer was replaced by a 50nm *n*-GaAs (8×10^{17} cm^{-3}) layer. The devices were grown by molecular beam epitaxy (MBE) on (100)-oriented n^+-GaAs substrates. The growth rate of the GaAs host material was 1.0μm / hr. Si and Be were used as *n*- and *p*-type dopants, respectively. The growth temperature of the GaAs material was 580°C. After finishing the growths,

111

circular mesas, 70μm in diameter, were made by etching down to the substrates for both devices. The mesa etchings were finished by employing $NH_4OH:H_2O_2:H_2O$ solution. The contacts to the cap layers of the mesas were made by using the evaporation and liftoff of AuGe in the form of a ring, which was enlarged on one side to enable bonding. The open areas for the optical windows had a diameter of 30μm. AuGe was also used as the ohmic contact metal to the substrates. Figure 1 shows the cross-sectional view of devices G220 and G330.

Results and Discussion

For an *npn* structure under a positive collector-to-emitter (C-E) voltage V_{CE} with base open, the collector current I_C is approximated to (13)

$$I_C = \frac{MI_{CO}}{1-\alpha M} \tag{1}$$

where

$$M = \frac{1}{1-\left(\dfrac{V_{CB}}{BV_{CBO}}\right)^n} \tag{2}$$

I_{CO} is the collector reverse saturation current including the base-collector (B-C) junction leakage current and photocurrent, α the common-base current gain, M the avalanche multiplication factor, BV_{CBO} the B-C breakdown voltage with emitter open, V_{CB} the external voltage drop across the B-C junction, and n a constant which is about 3.5 for GaAs (14). The switching performance in an *npn* structure operating in the avalanche region is produced by the positive feedback between the avalanche multiplication in the B-C junction and the current gain given by the transistor action. According to Eq.(2), M will increase with a positive V_{CE} voltage, since the reverse-biased B-C junction will absorb the most part of the voltage. The multiplied holes generated in the B-C junction will move toward the base, charge the E-B junction, and then reduce the E-B potential. More electrons are therefore injected from the emitter into the base and reach the B-C junction, which will generate more electron-hole pairs. This process is the so-called "positive feedback loop." For the V_{CE} voltage large enough to turn $1-\alpha M$ to zero, the B-C junction could breakdown, according to Eq.(1). From (2), the $1-\alpha M = 0$ condition can define a switching voltage V_S as

$$V_S = BV_{CBO}(1-\alpha)^{1/n}. \tag{3}$$

If α increases with I_C after $V_{CE} = V_S$, the criterion for the B-C breakdown will conduct a reduction in M for $M = 1/\alpha$. This M reduction requires a smaller B-C voltage. As a consequence, a negative-differential-resistance (NDR) phenomenon is expected in the breakdown characteristics. Then, a low-impedance ON state will be obtained if α saturates.

Figure 2 shows the band diagrams of devices G220 (solid lines) and G330 (dotted lines) under a positive V_{CE} voltage. As photons with energies higher than the band-gap energy of GaAs inject into the cap layer, electron-hole pairs will be photogenerated in both devices. The holes photogenerated in the n^+-GaAs layers on both sides of the devices could be neglected, as those holes will be recombined in the dense surrounding electron gas. In the emitter region, the photogenerated electrons will transport toward the base, and holes will move in opposite direction for the forward biasing in the E-B junction. In the base and collector regions, the photogenerated electron-hole pairs will be separated by the existing electric field. The photogenerated holes will move down to well A, and then be confined in well A or charge the E-B junction, as shown in figure 2. Owing to the present of the E-B valence-band discontinuity, the holes in well A will experience more effective confinement. On the other hand, the photogenerated electrons will move toward the substrate. For device G220, part of the electrons will be confined in well B, as shown in figure 2, due to the existence of the wide-gap AlGaAs layer in the collector region, and others will be collected by the electrode. The holes charging the E-B junction and the holes confined in well A will lower the potential barrier of the base for both devices. If the V_{CE} voltage is high enough, the breakdown will occur in the 200nm i-GaAs layer in the collector region. The multiplied holes will also be confined in well A or charge the E-B junction to lower the barrier, and then the electrons emitted over the barrier from the emitter to the base will be increased. This positive feedback process will result in an S-shaped NDR performance in the current-voltage (I-V) characteristics of both devices. Because the characteristics of both devices are influenced by the input-light power, the switching is controllable by changing the input-light. Because the photogenerated electrons confined in well B will increase the voltage drop across the 200nm i-GaAs layer in the collector region, the switching characteristics of device G220 will exhibit a higher optical sensitivity and a lower holding power than that of device G330, which will be discussed in the following.

The experimental I-V curves of devices G220 and G330 under dark and illumination at 300K are shown in figures 3 and 4, respectively. The illumination characteristics were obtained under a tungsten lamp as the light source. The incident optical power was measured from the photocurrents of the devices. It is estimated that the illumination power was 100nW. From figures 3 and 4, it is observed that with increasing illumination, the switching voltage V_S is reduced for both devices. This is attributed to the photogenerated carriers to increase the current gain, especially at low currents. At low collector currents, because the photogenerated holes charging the E-B junction and holes confined in well A will lower the potential barrier of the base, the current gain α will increase with the signal level. As α increases, the breakdown condition $M = 1/\alpha$ will appear at a lower value of M, which implies a lower V_{CE} voltage is sufficient to cause the breakdown. Thus the V_S under illumination is smaller than that under dark for both devices. The light input lowers the V_S about 1.9V for device G220 and about 0.9V for device G330, as shown in table 1. The higher optical sensitivity of device G220 is due to the photogenerated electrons confined in well B to increase the voltage drop across the 200nm i-GaAs layer in the collector region, which makes the V_S decrease more effectively. Under dark or illumination condition, the switching voltage V_S, switching current I_S, holding voltage V_H and holding current I_H of device G220 are all smaller than those of device G330, also as shown in table 1. At the OFF state of device G220 under a positive V_{CE} voltage, well B will confine electrons to increase the voltage drop across the collector region, and the wide-gap confinement layer will suppress the transporting electrons into the collector region. Thus the V_S and I_S of device G220 will be smaller. At the ON state of device G220, as the electrons confined in well B will give more positive polarity for the B-C junction and, therefore, elevate the band structure at the collector side, the V_H will be smaller in device G220. Because of the wide-gap confinement layer in the collector region to suppress the transporting electrons, the I_H of device G220 will also be smaller than that of device G330. Consequently, a lower holding power $P_H = I_H \times V_H$ is obtained in the G220 characteristics.

Conclusion

We have reported about the enhancement in optical sensitivity and the decrement in holding power of an *npn* optoelectronic switch by introducing a wide-gap confinement layer into the collector region. The performance improvement can result in a lower power consumption and larger fanin-to-fanout ratio, which are very useful for the design and fabrication of optical parallel processing.

Table 1. Switching parameters of devices G220 and G330 at 300K.

	(Device G220)		(Device G330)	
	Dark	Illuminated	Dark	Illuminated
Switching voltage, V_S	10.3V	8.4V	11.3V	10.4V
Switching current, I_S	1.6mA	4.2mA	3.9mA	8.0mA
Holding voltage, V_H	6.8V	6.3V	7.3V	6.9V
Holding current, I_H	6.2mA	5.0mA	12.9mA	10.7mA

Light

200nm	n^+-GaAs	$3\times10^{18}cm^{-3}$	(Cap)
200nm	N-Al$_{0.4}$Ga$_{0.6}$As	$5\times10^{17}cm^{-3}$	(Emitter)
80nm	p^+-GaAs	$8\times10^{18}cm^{-3}$	(Base)
50nm	n-GaAs	$8\times10^{17}cm^{-3}$	
50nm	N-Al$_{0.4}$Ga$_{0.6}$As (device G220) n-GaAs (device G330)	$8\times10^{17}cm^{-3}$	(Collector)
200nm	i-GaAs		
300nm	n^+-GaAs	$3\times10^{18}cm^{-3}$	(Buffer)
	n^+-GaAs Substrate		

Figure 1. Cross-sectional view of devices G220 and G330. The structure of device G330 is the same as that of device G220 except that the 50nm N-Al$_{0.4}$Ga$_{0.6}$As $(8\times10^{17}\,cm^{-3})$ layer is replaced by a 50nm n-GaAs $(8\times10^{17}\,cm^{-3})$ layer.

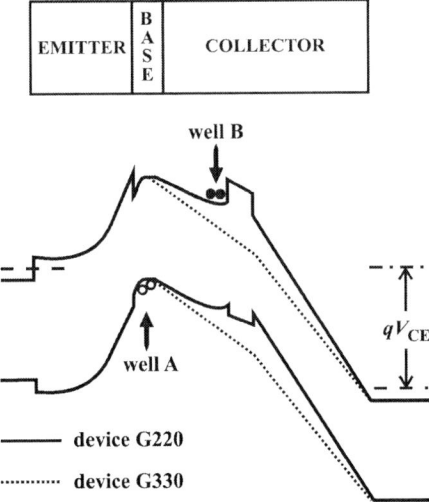

Figure 2. Band diagrams of devices G220 (solid lines) and G330 (dotted lines) under a positive V_{CE} voltage.

ECS Transactions, 28 (4) 111-118 (2010)

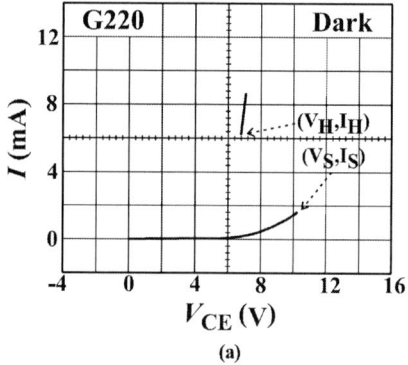

(a)

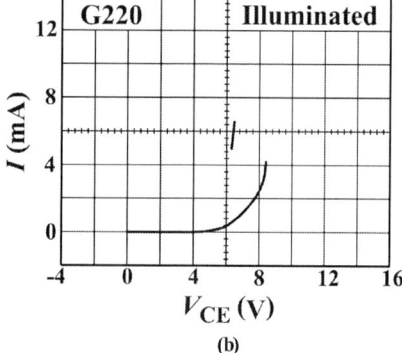

(b)

Figure 3.   Experimental current-voltage (*I-V*) curves of device G220 under (a) dark and (b) illumination at 300K.

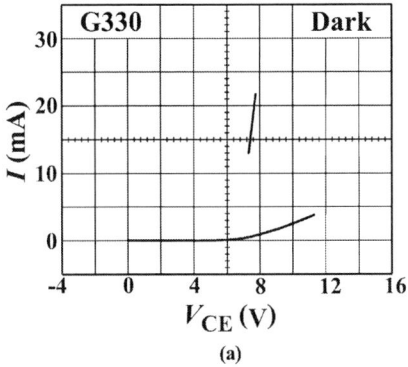

(a)

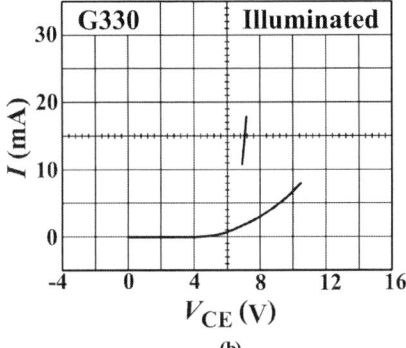

(b)

Figure 4. Experimental current-voltage (*I-V*) curves of device G330 under (a) dark and (b) illumination at 300K.

Acknowledgments

This work was supported by the National Science Council of the Republic of China under Contract No. NSC 98-2221-E-013-001.

References

1. H. Opper, J. Cai, R. B. Garber, R. Basilica and G. W. Taylor, *IEEE Trans. Electron Devices*, **51**, 1091 (2004).
2. G. W. Taylor, H. Opper, J. Cai, B. Garber and R. Basilica, *J. Appl. Phys.*, **96**, 7612 (2004).
3. D. F. Guo, *IEEE Electron Device Lett.*, **27**, 37 (2006).
4. S. J. Kovacic, B. J. Robinson, J. G. Simmons and D. A. Thompson, *IEEE Electron Device Lett.*, **14**, 54 (1993).
5. F. Y. Huang and H. Morkoc, *Appl. Phys. Lett.*, **64**, 405 (1994).
6. G. W. Taylor, D. L. Crawford and J. G. Simmons, *Appl. Phys. Lett.*, **54**, 543 (1989).
7. B. Kallback and H. Beneking, in *High-Speed Electronics*, 1986, 204.

8. K. Hara, K. Kojima, K. Mitsunaga and K. Kyuma, *IEEE J. Quantum Electronics*, **28**, 1335 (1992).
9. D. F. Guo, *IEE Proc.-Optoelectronics*, **148**, 121 (2001).
10. P. Papadopoulou, N. Georgoulas and A. Thanailakis, *Microelectron. J.*, **33**, 487 (2002).
11. S. B. Hwang, Y. K. Fang, K. H. Chen, C. R. Liu, J. D. Hwang, and M. H. Chou, *IEEE Trans. Electron Devices*, **40**, 721 (1993).
12. G. Cesare, G. Masini and F. Palma, *IEEE Trans. Electron Devices*, **43**, 1077 (1996).
13. S. M. Sze, *Physics of Semiconductor Devices*, Wiley, New York, 150(1981).
14. R. A. Logan, A. G. Chynoweth and B. G. Cohen, *Phys. Rev.*, **128**, 2518 (1962).

Drain Leakage Current in MuGFETs at High Temperatures

Jorge Giroldo Jr. and Marcello Bellodi

Department of Electrical Engineering, Centro Universitário da FEI, S. B. do Campo, São Paulo - 09850-901, Brazil

This paper will show the drain leakage current (I_{DLeak}) behavior in SOI Multiple-Gate devices (MuGFET) for double-gate (DGFinFET) and triple-gate (TGFinFET) configurations, operating since room temperature up to 300°C. Through three dimensional (3D) numerical simulations results is observed that I_{DLeak} is composed mainly by electrons for all devices operating at same conditions. Besides it, lower I_{DLeak} values are observed for all TGFinFET, when compared to DGFinFETs, analyzed in this investigation.

Introduction

Many applications such as automotive, industrial, aerospace, require the operation of integrated circuits at high temperatures (up to 300 °C), where MuGFET technology has shown good results for these applications in the future (1). At high temperatures, the main cause of failure observed in devices and circuits is due to the leakage current increases on drain/source to channel junctions (2). Multiple gates devices are the main promises to advance the limits of the integration scale due the best control of the channel (3-6).

This paper will discuss and show a comparison of I_{DLeak} behavior between DGFinFET and TGFinFET according to its channel length (L), width (W_{FIN}), height (H_{FIN}) for all devices operating since room temperature up to 300°C. In order to understand I_{DLeak} behavior as a function of these variables, it will be analyzed the drain leakage current carriers composition and their distribution along the width and height of the devices silicon film channel.

SOI MuGFET Structures

The devices characteristics used in all 3D simulations are Fin width (W_{FIN}) being 240, 120, 60 and 30 nm, Fin heights (H_{FIN}) are 240, 150, 120, 60 and 30 nm, the buried oxide (t_{BOX}) and gate oxide (t_{OX}) thickness are 145 nm and 2 nm, respectively. In order to avoid the top gate influences, DGFinFET has a gate oxide thickness (t_{OXsup}) of 100 nm. The channel length range analyzed was from 100 nm up to 1 μm. The channel region has a p-type doping concentration $N_A = 1 \times 10^{15}$ cm^{-3} while source and drain regions are a n-type which doping concentration is $N_D = 1 \times 10^{20}$ cm^{-3}, characterizing an n-channel MuGFETs.

Figure 1 show MuGFET structures of a DGFinFET (A) and TGFinFET (B), where it is possible to observe some details concerning to each device.

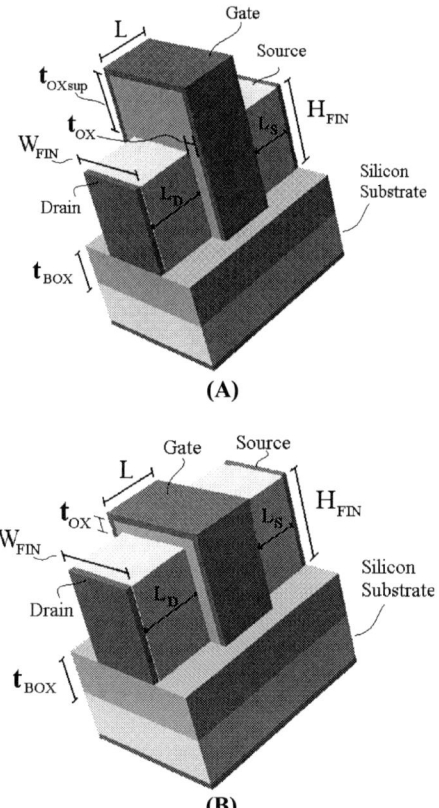

Figure 1. MuGFET structures of a DGFinFET (A) and a TGFinFET (B).

Three-Dimensional Numerical Simulations Results

The results presented along these investigations were obtained through 3D numerical simulations, performed with the numerical device simulator ATLAS (7).

In order to study I_{DLeak} it is necessary to obtain the drain current (I_{DS}) versus the gate voltage (V_{GS}) curves, for all temperature range analyzed. I_{DLeak} values were extracted for all devices operating at a constant drain voltage $V_{DS}= 25$ mV.

Figure 2 shows some results concerned to I_{DS} x V_{GS} curves, since room temperature up to 300°C, for a DGFinFET (A) and TGFinFET (B) for L = 500 nm while W_{FIN} and H_{FIN} assume 120 nm. From these results is possible to note that as the temperatures increases, the subthreshold slope decreases, showing that I_{DLeak} increases with the temperature. Besides it, in the triode region ($V_{GS} > 0.5$ V), I_{DS} reduces due to the mobility reduction as the temperature rises. Similar results were observed for all L range and for all combinations between H_{FIN} and W_{FIN} analyzed along this paper.

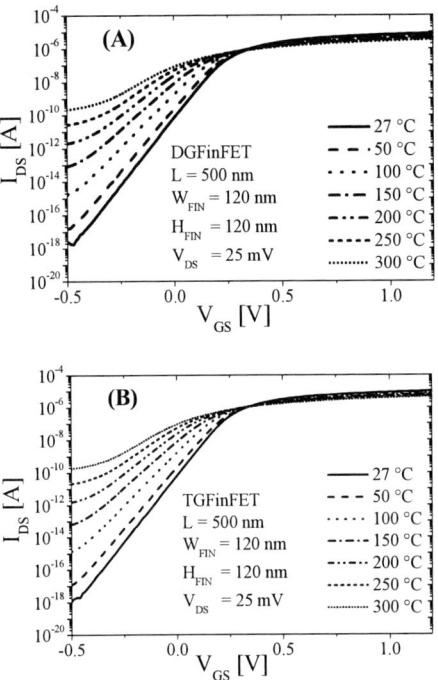

Figure 2. I_{DS} x V_{GS} curves for DGFinFET (A) and TGFinFET (B), for L= 500 nm and $W_{FIN}=H_{FIN}$ = 120 nm, as a function of the temperature.

In order to extract the total drain leakage current I_{DLeak}, it was necessary to bias all devices at a same V_{GS} to guarantee that they are operating in the leakage region. Then, after analyzing all I_{DS} x V_{GS} curves, it was imposed that I_{DLeak} can be extracted for V_{GS} being -0.5 V, where I_{DLeak} tends to be almost constant at the same temperature of operation.

Some results are shown in Figure 3 for I_{DLeak} behavior as a function of the temperature and the channel length. Figure 3(A) shows that I_{DLeak} increases as the temperature rises in all devices evaluated. Note that when L= 1 μm, I_{DLeak} assumes the lowest values for all temperature range, but when the devices are submitted to high temperatures, a greater variation in I_{DLeak} levels is observed. When I_{DLeak} is analyzed as a function of L, as shown in Figure 3(B), it is possible to observe that at same temperature, I_{DLeak} increases significantly as L reduces. It is important to see that TGFinFET showed lower I_{DLeak} values for all temperature range studied, when compared to DGFinFETs operating at same conditions.

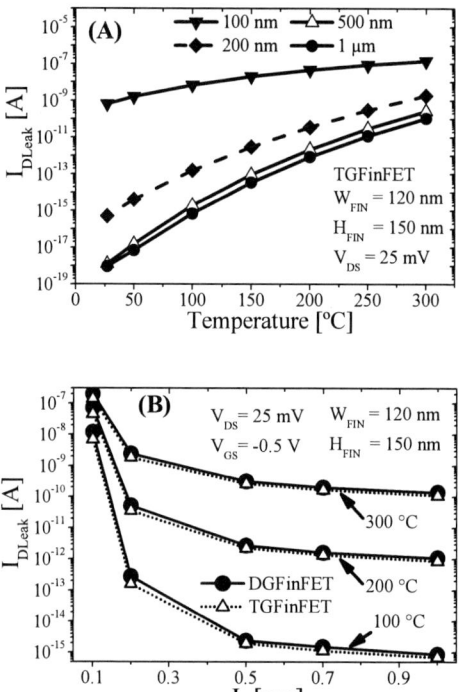

Figure 3. I_{DLeak} behavior as a function of the temperature (A) and channel length (B).

The influence of the film silicon width (W_{FIN}) on I_{DLeak} is presented in Figure 4. In order to perform this kind of analysis it was used a constant value for H_{FIN}= 120 nm for all devices analyzed, while L and W_{FIN} have changed. In figure 4 (A) it can be seen I_{DLeak} behavior for a DGFinFET and in the Figure 4 (B) for a TGFinFET, both with L= 1 µm and as a function of the temperature of operation.

The results shown that W_{FIN} increase contributes to I_{DLeak} rises for all temperature range evaluated. Besides it, it can be seen that I_{DLeak} levels are one order of magnitude lower in TGFinFET devices due to the top gate influences on the channel control.

On the other hand, as L reduces, similar I_{DLeak} behavior concerning W_{FIN} variation was observed in all devices evaluated in this investigation.

Figures 4 (C) and 4 (D) show that for L= 200 nm, I_{DLeak} also becomes higher as W_{FIN} has its value increased for all temperature range evaluated.

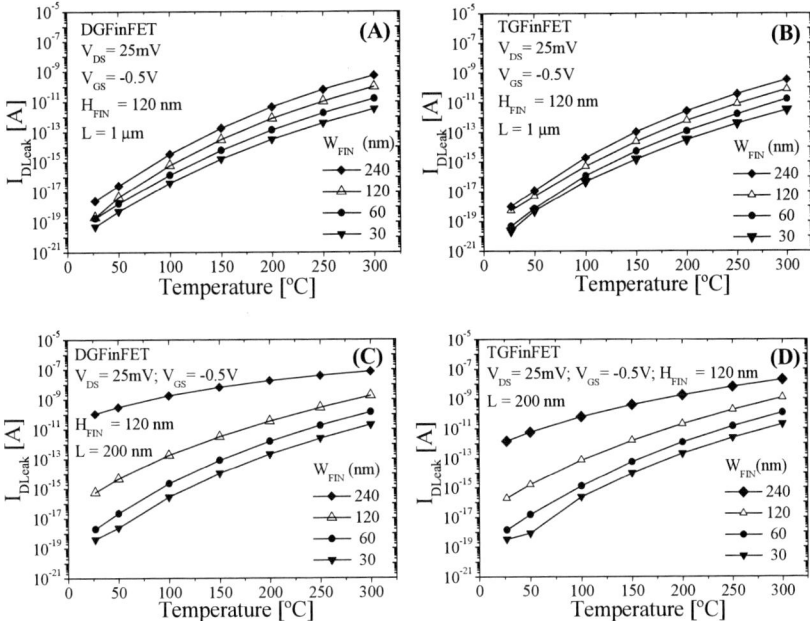

Figure 4. DGFinFET and TGFinFET I_{DLeak} behavior as a function of the temperature for some W_{FIN} values for L= 1 µm (A,B), and for L= 200 nm (C, D), respectively.

Another investigation done was the height variation (H_{FIN}) influences in I_{DLeak} behavior, where W_{FIN} being a constant value of 120 nm and for H_{FIN} and L assuming some different values. These results are shown in Figure 5.

As it can be seen in Figure 5, H_{FIN} increase contributes significantly to grow up I_{DLeak}. Again, the results show that for all TGFinFET devices presented a significantly lower I_{DLeak} when compared to DGFinFET, operating at same conditions.

Then, analyzing these results, it can be seen that as L, H_{FIN} and W_{FIN} assume new values, I_{DLeak} also changes. Then, it indicates that I_{DLeak} is proportional to the devices junctions areas (drain/channel and channel/source), assuming higher values when these areas are increased by H_{FIN} or W_{FIN}.

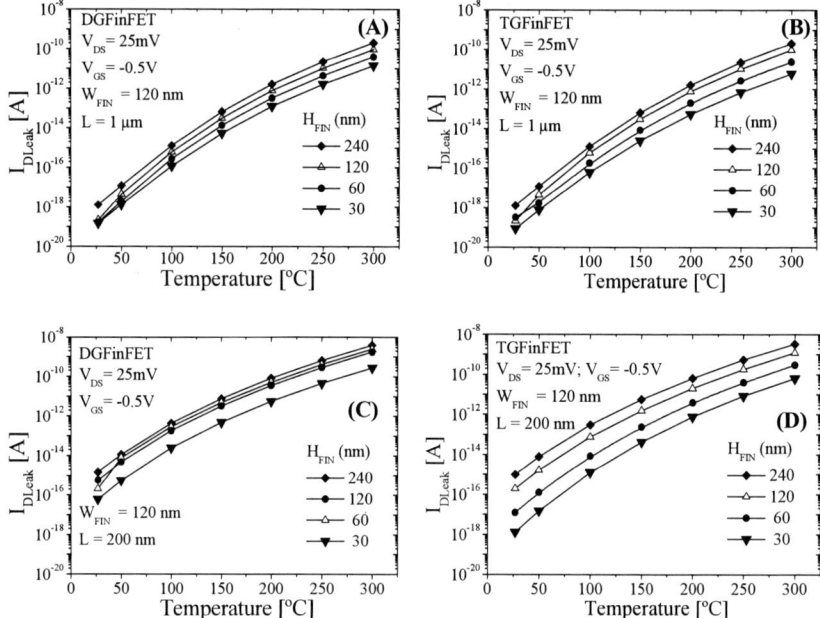

Figure 5. DGFinFET and TGFinFET I_{DLeak} behavior as a function of the temperature for some H_{FIN} values for L= 1 μm (A, B), and for L= 200 nm (C, D), respectively.

In order to better understand I_{DLeak} behavior of these devices under investigation, the total drain leakage current density (J_{TLeak}) as well as its composition (electrons and holes) was carefully investigated. The densities have been extracted through W_{FIN} and H_{FIN} in order to understand how each device gate influences in the I_{DLeak} behavior.

Figure 6 shows the total drain leakage current density (J_{TLeak}) and its distribution along the silicon film channel. In Figure 6 (A), J_{TLeak} distribution is showed through W_{FIN} in a DGFinFET, where it can be seen that it flows predominantly into the middle of the silicon film volume and its density reduces near the gate interfaces. Similar behavior also was observed in a TGFinFET as shown in Figure 6 (B). For both cases, as the temperature rises, the same distribution is observed for all devices but the drain leakage current density levels are higher than the one found in a device when it is operating at lower temperatures.

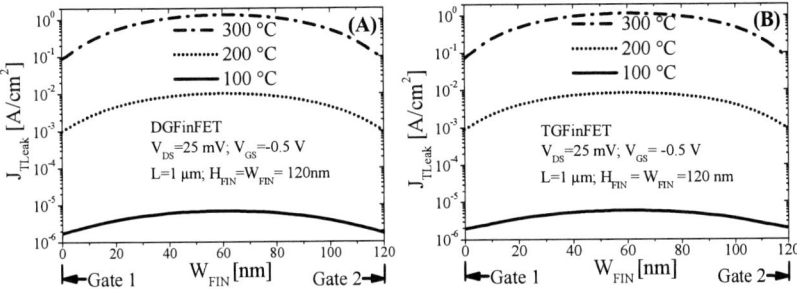

Figure 6: J_{TLeak} as a function of W_{FIN} for DGFinFET(A) and TGFinFET (B), at high temperatures.

Similar results for J_{TLeak} distribution through W_{FIN} were found in all DGFinFET and TGFinFET when W_{FIN} assumes new values.

Now, analyzing J_{TLeak} behavior as a function of H_{FIN}, it can be seen that for DGFinFET (Figure 7(A)) J_{TLeak} is almost constant along H_{FIN}. On the other hand, for TGFinFET as showed in Figure 7(B), J_{TLeak} reduces significantly near the top gate interface, for the devices with L = 1μm.

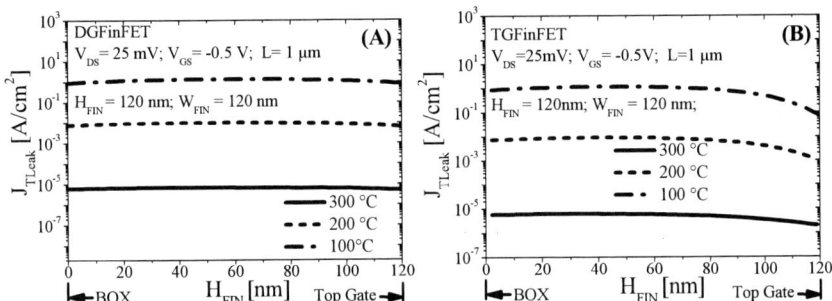

Figure 7. J_{TLeak} as a function of H_{FIN} for DGFinFET (A) and TGFinFET (B), operating at high temperatures.

Similar behavior was also observed for MuGFETs when the channel length is 500 nm, as presented in Figure 8. But in this case, for H_{FIN}= 150 nm, it naturally causes an I_{DLeak} increases, due to the junctions areas become larger.

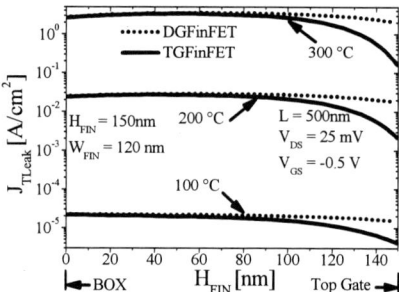

Figure 8. Comparison of J_{TLeak} as a function of H_{FIN} for DGFinFET and TGFinFET for L= 500 nm, at high temperatures.

Figure 9 shows a comparison between a DGFinFET and TGFinFET for H_{FIN}= 30 nm, where it is possible to see that J_{TLeak} is lower when compared to devices with larger H_{FIN}, as shown previously. From these results, also it can be seen that the TGFinFET top gate has influence through the height of the silicon film more significantly, while in DGFinFET J_{TLeak} is practically constant through H_{FIN}, similar results are observed for lower L.

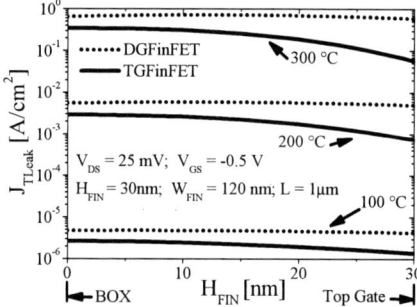

Figure 9. Comparison of J_{TLeak}, as a function of H_{FIN}, between DGFinFET and TGFinFET for L= 1 µm, W_{FIN}= 120 nm and H_{FIN}= 30 nm, at high temperatures.

It is important to mention that for all TGFinFET investigated along this investigation; it was observed the same reduction in drain leakage current density in the areas near the top gate, independent of H_{FIN} values.

Now, analyzing the drain leakage current composition (carries – holes and electrons), as shown in Figure 10 (A) for a DGFinFET which L= 1 µm, at 300°C, it is possible to observed that the drain leakage current flows mainly in the middle of the silicon film volume, being predominantly composed by electrons ($J_{Electrons}$). Note in this Figure that the holes concentration (J_{Holes}) is more significant in the gate interfaces, due to the negative bias applied to the gate (V_{GS} = -0.5 V) which naturally leads to the accumulation of holes in this region of the devices.

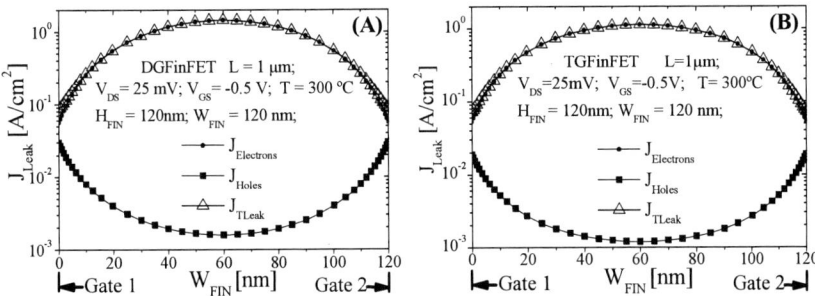

Figure 10. J_{TLeak} composition as a function of W_{FIN} for a DGFinFET(A) and TGFinFET(B), at T= 300°C.

Meanwhile, in the middle of the width W_{FIN}, there is a difference of almost three orders of magnitude between the densities of electrons and holes (10^0 versus 10^{-3}, approximately). Analyzing a TGFinFET at same conditions of temperature and bias and for the same channel length, it was observed that the drain leakage current density and its distribution along the silicon film are similar, as shown in Figure 10 (B) but its total density is slightly lower.

According to the results presented in Figure 11 (A), it can be seen that the drain leakage current density of a DGFinFET along the H_{FIN}, is mainly composed by electrons and its intensity around the top gate interface tends to reduces.

For TGFinFET, as shown in Figure 11 (B), the drain leakage current also is composed mainly by electrons across the height H_{FIN}, however in the top gate interface there is a significant reduction of one order of magnitude in the overall drain leakage current density and also, high densities of holes, since the top gate is biased with a negative value (V_{GS} = -0.5 V), which allows the accumulation of holes and consequently, reducing I_{DLeak}.

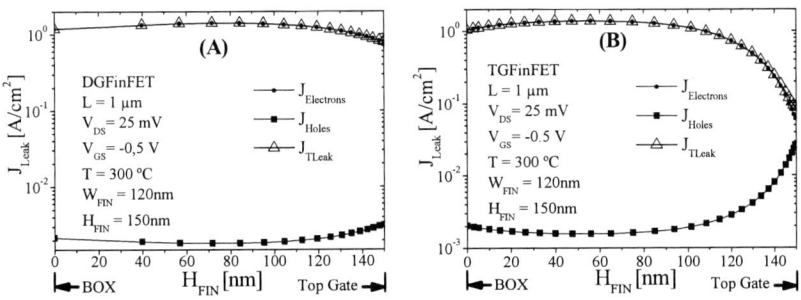

Figure 11. Distribution of the total drain leakage current density along H_{FIN} for DGFinFET (A) and TGFinFET (B), for L= 1 μm at 300°C.

Similar results were obtained by changing W_{FIN} and H_{FIN} for all temperature and L ranges analyzed along this work. Figure 12 shows some results for DGFinFET (A) and TGFinFET (B), for L= 1 µm, H_{FIN}= 30 nm and W_{FIN}= 120 nm.

Figure 12 (A) shows the drain leakage current density on a DGFinFET obtained as a function of W_{FIN} while Figure 12 (B), shows the results obtained for H_{FIN} for the same device, it can be seen similarity of J_{TLeak} distribution and its composition remains composed mainly by electrons.

Looking at figure 12 (C) for a TGFinFET, also observed the same characteristics of J_{TLeak} distribution through W_{FIN}, however we see that the most significant J_{TLeak} attenuation in the center of silicon film. Figure 12 (D) shows the behavior of J_{TLeak} through H_{FIN} in which it is observed that the top gate is capable of influencing the control channel of the device since its interface to a little more than half of H_{FIN}, contributing to the reduction of I_{DLeak}.

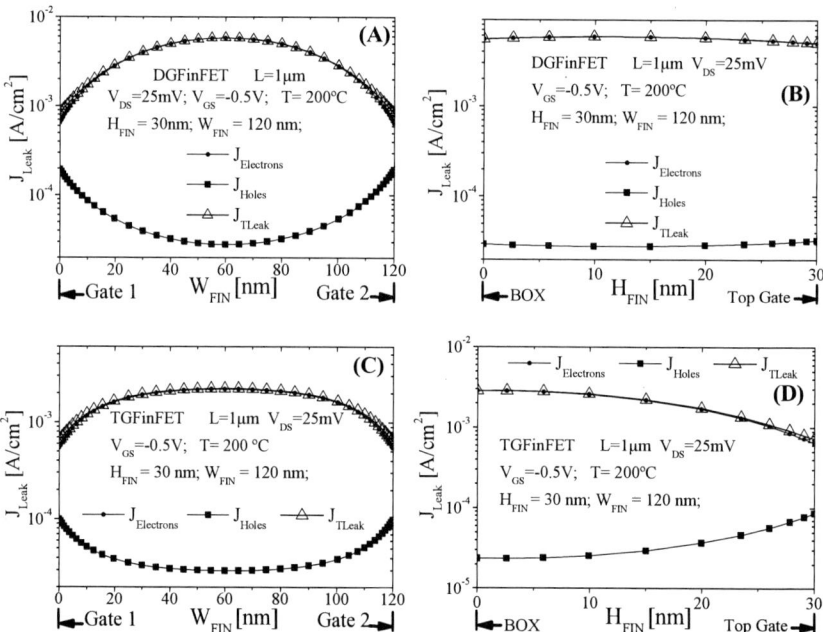

Figure 12. Distribution J_{TLeak} in W_{FIN} and H_{FIN} to DGFinFET (A,B) and TGFINFET (C,D), at 200°C.

In figure 13 it can be note that the reduction of L to 100 nm causes a rise in J_{TLeak} levels, however their distribution along the silicon film is still similar along W_{FIN} (A) and H_{FIN} (B) and their composition shows that J_{Leak} is mainly composed by electrons. In this situation, a reduction of 3 orders of magnitude is observed due to the top gate.

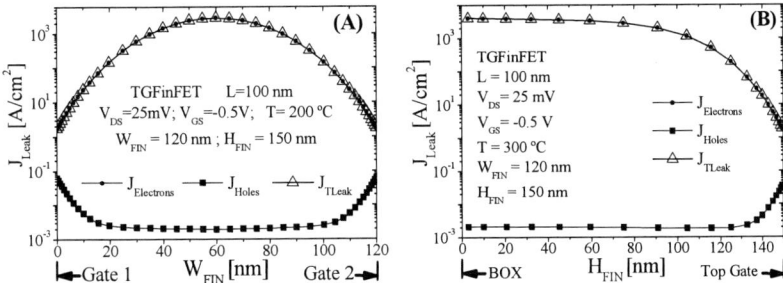

Figure 13: Distribution J_{TLeak} as a function of W_{FIN} (A) and H_{FIN} (B).

Conclusions

This paper presented the total drain leakage current behavior of SOI nMOSFETs DGFinFET and TGFinFET operating since room temperature (27°C) up to 300°C. Through three-dimensional numerical simulations results it was observed that I_{DLeak} increases when the devices are submitted at high temperatures and shows higher I_{DLeak} values as L decreases.

For all the devices analyzed, rising W_{FIN} or H_{FIN} values, it was noticed a significant contribution to I_{DLeak} increase, due to the drain / channel and channel / source junctions areas increase.

The total drain leakage current composition is given mainly by electrons which current flows through the silicon film thickness, for both SOI MOSFETs. Once the top gate in TGFinFET contributes strongly to I_{DLeak} reduction, then this explain a smaller I_{DLeak} observed in these devices when compared to the DGFinFET operating at the same bias and temperature conditions.

References

1. J.P. Colinge, FinFETs and Other Multi-Gate Transistors, Springer (2008).
2. M. Bellodi, J.A. Martino in Solid-State Elec. 45, pp 683-688 (2001).
3. V. Kilchytska, N. Collaert, M. Jurczak, D. Flandre in Solid-State Elec. 51 pp 1185–1193 (2007).
4. D. Schimitt-Landsiedel, C. Werner in Solid-State Elec. 53 pp 411–417 (2009).
5. Chi-Woo Lee, Se-Re-Na Yun, Chong-Gun Yu, Jong-Tae Park , Jean-Pierre Colinge, in Solid-State Electronics 51 pp 505–510, (2007).
6. T. Poiroux, M. Vinet, O. Faynot, J. Widiez, J. Lolivier, T. Ernst, B. Previtali, S. Deleonibus in Microelectronic Engineering 80 pp 378–385 (2005).
7. ATLAS device simulation, v. 5 v. 5.14.0.R, Silvaco Intern. (2008).

ECS Transactions, 28 (4) 131-135 (2010)
10.1149/1.3377109 ©The Electrochemical Society

Fabrication of IGZO Sputtering Target and Its Applications to the Preparation of Thin-film Transistor (TFT) Devices

Chun-Chieh Lo and Tsung-Eong Hsieh*

Department of Materials Science and Engineering, National Chiao Tung University,
1001 Ta-Hsueh Road, Hsinchu, Taiwan 30010, R.O.C.

A hybrid of chemical dispersion and mechanical grinding process was developed to fabricate the mixture of nano-scale In_2O_3, Ga_2O_3 and ZnO oxide powders at the atomic ratio 1:1:2. As revealed by x-ray diffraction (XRD) analysis, sputtering target containing sole $IGZO_4$ phase could be obtained by sintering the oxide mixture pressed in disc form at temperatures $\geq 1300°C$ for 6 hrs. The IGZO target was then transferred to a sputtering system and the thin-film transistors (TFTs) containing amorphous IGZO channels were fabricated. Post-annealing at $300°C$ for 1 hr in air ambient was performed in order to improve the device performance. Electrical measurements indicated that the TFT samples with saturation mobility (μ_{sat}) = 14.7 $cm^2/V·s$, threshold voltage (V_{TH}) = 0.57 V, subthreshold swing (S.S.) = 0.45 V/decade and on/off ratio = 10^8 could be achieved.

Introduction

Transparent amorphous oxide semiconductors (AOSs) are promising materials for channel layer of thin-film transistors (TFTs) because AOS TFTs exhibit a relatively high saturation mobility (μ_{sat} >10 $cm^2/V·s$), high film flatness, low processing temperatures and high optical transparency in the visible region [1]. However, some AOS materials have uncontrollable carriers generated due to the presence of oxygen vacancies. Therefore, it is crucial to develop an appropriate AOS material with satisfactory physical properties and stable charge carrier feature to fulfill practical applications. Recently, amorphous In-Ga-Zn-O (a-IGZO) thin films for channel layer of transparent TFTs have attracted a lot of attentions due to their unique physical properties and processing advantages [2]. The a-IGZO channel layer prepared *via* sputtering deposition process demonstrates its feasibility to large-area TFT device fabrication [3]. The dependence of the TFT characteristics on the elemental composition of IGZO has been investigated in detail [4]. In this study, we reported the preparation of sputtering target containing sole $IGZO_4$ phase by using a hybrid of chemical dispersion and mechanical grinding process. The self-made IGZO target was then utilized to fabricate the TFT devices on Si and glass substrates, respectively, and key electrical properties were measured. The deposited IGZO films was characterized by x-ray diffraction (XRD), x-ray photoelectron spectroscopy (XPS), atomic force microscopy (AFM) and UV-visible spectrometer so as to elucidate the relationship between film microstructure and TFT device performance.

Experimental

For the preparation of precursor oxide powders for IGZO target, as-received ZnO (purity = 99.999%, Seedchem/Australia) Ga_2O_3 (purity = 99.995%, ELECMAT/USA) and In_2O_3 (purity = 99.99%, CERAC/USA) powders were first rinsed in ethanol. The

In_2O_3, Ga_2O_3 and ZnO powders were mixed at the molar ratio = 1:1:2 and a hybrid of chemical dispersion and mechanical grinding process [5] was then carried out to fabricate the aqueous suspension containing nano-scale In_2O_3, Ga_2O_3 and ZnO oxide powder mixture. After drying, the powder mixture was pressed into disc form and consequently sintered at temperatures ranging from 900 to 1400°C for various time spans so as to identify the condition for $IGZO_4$ phase formation. The self-made IGZO target was transferred into a sputtering system and the TFT devices containing 15-nm thick amorphous IGZO channels were fabricated on n^+-Si substrate containing 200-nm thick thermal oxide layer and glass substrates, respectively. The sputtering was carried out at a Ar/O_2 gas flow ratio = 20/1.2, RF power = 80W and working pressure = 1 mtorr. 300-nm thick aluminum (Al) metal lines deposited by e-beam evaporation served as the source and drain electrodes of devices. Finally, the annealing treatments at temperatures of 300°C for 1 hr in air ambient were performed in order to improve the device performance.

Results and Discussion

By sintering the In_2O_3, Ga_2O_3 and ZnO powder mixtures at temperatures ranging from 900 to 1400°C for various times, it was found that the sole $IGZO_4$ phase could be obtained via a sintering at temperatures $\geq$ 1300°C for 6 hrs. Figure 1 presents the XRD patterns of the powder mixtures before and after the 1300°C/6-hr sintering treatment and it illustrates the presence of $IGZO_4$ phase in the self-made target in accord with the JCPDS-381104 standard.

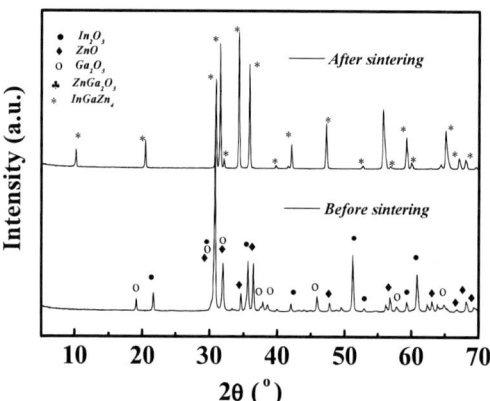

Figure 1. XRD patterns of In_2O_3, Ga_2O_3 and ZnO powder mixture before and after the 1300°C/6-hr sintering treatment.

Sputtering deposition was performed by using the self-made IGZO target and the XRD pattern of a 100-nm thick IGZO film subjected to a heat treatment at 300°C for 1 hr is shown in Fig. 2. It indicates the IGZO film remains amorphous even though it has been heated in such a high-temperature ambient. A cross-sectional TEM (XTEM) micrograph of 15-nm thick IGZO channel layer in TFT sample is displayed in Fig. 3. The selected area electron diffraction (SAED) pattern taken from the vicinity of IGZO region shows vague diffraction rings, again illustrating the amorphous feature of IGZO layer. Further, the IGZO film exhibits a very smooth surface with an average roughness ≈ 0.15 nm as revealed by AFM characterization.

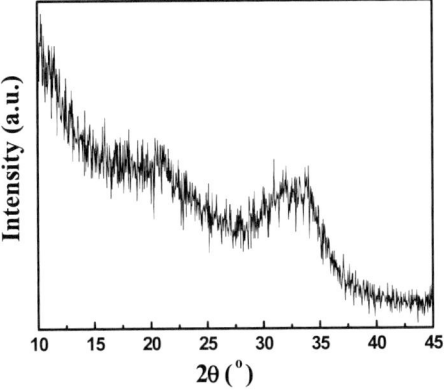

Figure 2. XRD of IGZO films deposited by using the self-made IGZO target. The film is about 100-nm thick and is subjected to a heat treatment at 300°C for 1 hr.

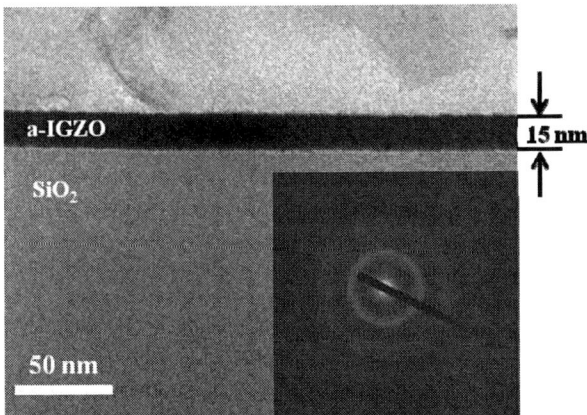

Figure 3. XTEM micrograph of 15-nm thick IGZO channel in a TFT sample made on Si substrate.

Figure 4 shows the transmittance of IGZO film subjected to the 300°C/1-hr post-annealing. It indicates that the film possesses an average transmittance ≥ 85% in the visible-light wavelength range, illustrating its high optical transparency. There are two crucial properties of transparent conducting oxides (TCOs), e.g., transparency and conductivity. The result shown in Fig. 4 hence implies the feasibility of IGZO films for fully transparent TFT fabrication.

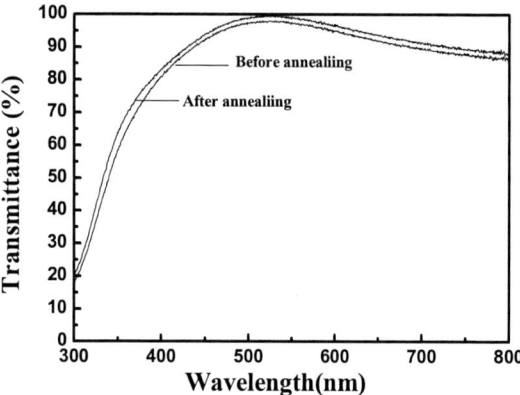

Figure 4. Transmittance of IGZO film subjected to the 300°C/1-hr post annealing.

Electrical characteristics of TFT samples were measured by using a semiconductor parameter measurement system containing a Keithley 4200 *I-V* source meter. The saturation mobility (μ_{sat}) and threshold voltage (V_{th}) were derived from a linear fitting of $\sqrt{I_{DS}}$ *versus* V_{GS} plot (I_{DS} = source to drain current; V_{GS} = gate to source voltage) using the equation

$$I_{DS} = \left(\frac{\mu_{sat} W \varepsilon_o \varepsilon_r}{2Ld} \right)(V_{GS} - V_{th})^2$$

where ε_o = dielectric constant of vacuum, ε_r = relative dielectric constant of the gate insulator (= 3.9 for SiO$_2$), L = device channel length, W = device channel width and d = the gate insulator thickness [6]. In this work, $L = 0.1$ mm and $W = 1.6$ mm, respectively.

Figure 5 shows the transfer characteristics of the a-IGZO TFTs before and after 300°C/1-hr post annealing at the drain to source voltages (V_{DS}) ranging from 5V to 15V. It can be seen that the TFT sample without post-annealing possesses μ_{sat}= 1.2 cm^2/V·s, V_{th} = 1.92 V, subthreshold swing (S.S.) = 0.85 V/decade and on/off ratio = 5×10^4. Post annealing dramatically improved TFT device performance that μ_{sat} = 14.7 cm^2/V·s, V_{th} = 0.57 V, S.S. = 0.45 V/decade and on/off ratio = 10^8 were observed. Analytical results presented above illustrate not only the good electrical performance of TFT devices prepared by using the self-made IGZO sputtering target, but also the benefits of post annealing on property improvement. It is believed that the post-annealing treatment remedies the oxygen-deficiency and amplifies the semiconductor features of IGZO layers so as to improve the electrical properties.

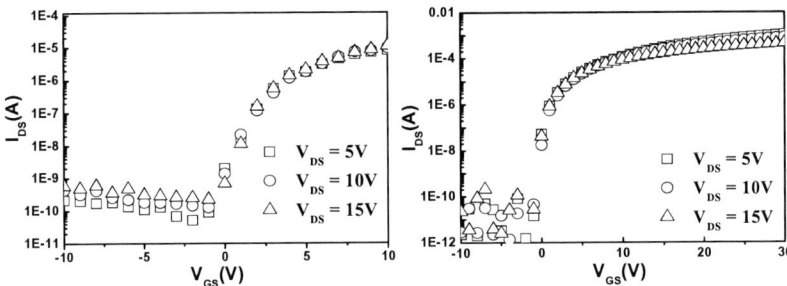

Figure 5. Transfer characteristics of a-IGZO TFTs (a) before and (b) after the post annealing at 300°C for 1 hr (V_{DS} = 5V to 15V).

In summary, mixture of nano-scale In_2O_3, Ga_2O_3 and ZnO oxide powders was prepared by a hybrid of chemical dispersion and mechanical grinding process. The powder mixture was then pressed and sintered at 1300°C for 6 hrs to form the sputtering target containing sole $IGZO_4$ phase. The target was utilized to deposit the channel layer of TFTs and XRD and TEM characterizations revealed that the amorphous IGZO films can be successfully made by using the self-made sputtering target. The a-IGZO layer possesses high optical transmittance ($\geq$ 85%) and is feasible to fully transparent TFT fabrication. Electrical property measurement showed that the TFT device subjected to 300°C/1-hr post annealing exhibits good transfer characteristics with μ_{sat}= 14.7 cm^2/V·s, V_{th} = 0.57 V, S.S. = 0.45 V/decade and on/off ratio = 10^8.

Acknowledgement

This work was supported by Chunghwa Picture Tubes. Inc. (CPT) project under the contract No. 97C173. Part of financial aid supported by National Science Council (NSC), Taiwan, R.O.C., under the contract No. NSC97-2221-E-009-029-MY3 is also deeply acknowledged.

References

1. H. Hosono, *J. Non-Cryst.Sol*, 352, 851-858 (2006).
2. K. Nomura, H. Ohta, A. Takaji, T. Kamlya, M. Hirano, H. Hosono, *Nature* 432 488 (2004)
3. K. Nomura, A. Takagi, T. Kamiya, H. Ohta, M. Hirano, and H. Hosono, *Jpn. J. Appl. Phys.* 45, 4303-4308 (2006).
4. N. J. DiNardo, in *Metallized Plastics 1*, K. L. Mittal and J. R. Susko, Editors, p. 137, Plenum Press, New York (1989).
5. K-L Ying, T-E Hsieh and Y-F Hsieh, *Ceramics International*, 35, 1165-1171 (2009).
6. *Thin-film Transistor*, edited by C. R. Kagan and P. Andry (Dekker, New York, 2003), p. 38.

Low-Resistivity and High-Transmittance Indium Gallium Zinc Oxide Films Prepared by Co-Sputtering In$_2$Ga$_2$ZnO$_7$ and In$_2$O$_3$ Targets

H. J. Chang[a], K. M. Huang[a], C. H. Chu[b], S. F. Chen[a], T. H. Huang[a], and M. C. Wu[a]

[a] Institute of Electronics Engineering, National Tsing Hua University, Hsinchu, 30013, Taiwan
[b] Institute of Photonics Technologies, National Tsing Hua University, Hsinchu, 30013, Taiwan

Indium gallium zinc oxide (IGZO) films are deposited on glass substrates by co-sputtering In$_2$Ga$_2$ZnO$_7$ and In$_2$O$_3$ targets. The structural, electrical, and optical properties of the IGZO films have been discussed as functions of substrate temperature and rf power supplied to the In$_2$O$_3$ target (P$_{rf,\ In2O3}$). From x-ray diffraction patterns, a halo peak around 34° is observed, which is attributed to the amorphous-like Zn$_k$In$_2$O$_{3+k}$ structure. The electrical resistivity decreases with increasing substrate temperature and P$_{rf,\ In2O3}$. The optimum resistivity is 7.16×10^{-4} Ω-cm obtained at 175 °C substrate temperature and 20 W P$_{rf,In2O3}$. The optical characteristics indicate that IGZO films are direct-transition type semiconductors. The optical band gap energy widens with increasing In$_2$O$_3$ content in the IGZO films. The optical transmittance of films deposited with 10-W P$_{rf,\ In2O3}$ is higher than 99% in the wavelength region from 516 to 535 nm.

1. Introduction

In recent years, transparent conducting oxides (TCOs) have attracted much of attention because of the highly demand for optoelectronic devices such as light-emitting diodes (LEDs), organic light-emitting diodes (OLEDs), and solar cells (1-5). Commercially used TCOs have the characteristics of high electrical conductivity and high optical transparency in the visible region (>80%). Nowadays, indium tin oxide (ITO) is the most commonly used TCO material, and numerous studies on its properties have been published (6-9). However, several drawbacks limit its applications. For example, transparency of ITO films in blue and near-UV region decreases rapidly, and they show Schottky behavior even after thermal annealing (10). Besides the above, the cost of ITO fabrication remains high.

In contrast to the disadvantages of ITO, zinc-oxide (ZnO)-based materials such as ZnO, indium zinc oxide (IZO), aluminum zinc oxide (AZO), and indium gallium zinc oxide (IGZO) are the newest alternatives. They form better alternatives due to the high carrier mobility, superior chemical selectivity, excellent environmental stability, and an absence of toxicity (11-13). Among them, IGZO is especially a new candidate used on optoelectronic devices due to its high transmittance, excellent surface smoothness, and low processing temperature. However, the electrical resistivity of undoped-IGZO is still high for it to be used for transparent electrode applications. In 2005, Takagi *et al.* discussed the effect of oxygen partial pressure during IGZO deposition (14). In 2008, Wantae *et al.* reported that the electrical resistivity of IGZO films was around 10^{-1} Ω-cm, produced by optimizing the deposition power in the sputtering process (15). Other works have focused on the relationship between stoichiometry and electrical properties (16-17).

However, a systematic discussion on structural, optical, and electrical properties of IGZO has never been truly formulated, and little attention has been paid to the process parameters.

Various deposition methods have been developed to prepare the IGZO films, including sputtering (15, 18-19), pulsed laser deposition (PLD) (20-21), and solution process (22). Among these, sputtering is the most preferable process, because it allows easy process control and the production with large areas. In this paper, we fabricated IGZO films by using $In_2Ga_2ZnO_7$ and In_2O_3 targets co-sputtered. The relationship between deposition parameters and film properties has been investigated systematically. It was observed that electrical resistivity of deposited IGZO films decreased as deposition temperature increased, and reached the minimum of 3.1×10^{-2} Ω-cm at $175°C$. Through XRD examination and reported data for on-set crystallization temperature, the deposited IGZO films were amorphous. Thus, the main reason for resistivity decrease is due to thermally generated oxygen vacancies. By using $In_2Ga_2ZnO_7$ and In_2O_3 targets co-sputtered, we raised the amount of In content in IGZO films, and further decreased the resistivity to 7.16×10^{-4} Ω-cm. From the optical transmittance and electrical resistivity data reported in this paper, the IGZO films are highlighted to be alluring candidates as transparent electrodes for optoelectronic devices.

2. Experimental

IGZO films were deposited on glass substrates (20×20 mm^2 area). Prior to film deposition, glass substrates were ultrasonically degreased using acetone, isopropyl alcohol, and deionized (DI) water for 10 min during each step, followed by nitrogen blowing. Film deposition was carried out using rf magnetron sputtering. 2-inch $In_2Ga_2ZnO_7$ (purity 99.99% with In_2O_3:Ga_2O_3:ZnO = 1:1:1 (at%)) and In_2O_3 (purity 99.99%) targets were employed as sputtering sources. The base pressure of the deposition chamber was $\sim 10^{-6}$ mTorr, and the working pressure was maintained at 5 mTorr under pure argon ambient. Although it is not common to sputter in pure argon ambient because the film composition may vary from the target composition, we have already investigated that it provides better long-term stability for the IGZO films. The glass substrates were placed on a stainless steel sample holder with a substrate-to-target distance of 10 cm. The substrate temperature was varied from room temperature to 275 $°C$ during the sputtering process, as determined from the temperature indicators attached to the sample holder. The rf power supplied to the $In_2Ga_2ZnO_7$ target was fixed at 50 W, while the DC bias was around 173 V. And, the rf power supplied to the In_2O_3 target was varied from 0 to 20 W with a DC bias around 110 V. Prior to film deposition, a pre-sputtering procedure amounting to 5 min was conducted in order to clean the target surface.

The thickness of the deposited IGZO films was between 100 to 200 nm, as measured by surface profiler (Tencor, Alpha-Step 500) and cross-section scanning electron microscopy (SEM). IGZO films with thickness ~ 200 nm were used for grazing incident x-ray diffraction (GIXRD, Rigaku TTRAX) analysis to investigate film structure and crystallinity. A Ni-filtered Cu Kα (λ = 1.5406 Å) source was equipped on the GIXRD system. The scanning range was between $2\theta = 5$ and $90°$, and the scanning rate was $2°$ per min. The electrical resistivity was measured using a four-point probe (Napson, RT-70). Optical transmittance spectra were measured by employing an ultraviolet-visible-near infrared (UV-Vis-NIR) spectrophotometer (Hitachi, U3010), over wavelength ranging from 300 to 800 nm. The transmittance was automatically calibrated against a bare glass as a reference sample.

3. Results and Discussion

3.1 IGZO Films Prepared by Using Single $In_2Ga_2ZnO_7$ Target

3.1.1 Growth Rate

In order to control the deposition process and the quality of IGZO films precisely, it is essential to know the relationship between sputtering parameters and growth rate. Therefore we investigated how the deposition parameters, such as rf power supplied to targets (P_{rf}) and substrate temperature (T_s), affect growth rate. Figure 1 shows the variation of the growth rate with respect to P_{rf} that supplied to $In_2Ga_2ZnO_7$ target. As Ts was fixed at 25 ℃ and P_{rf} was raised from 20 to 70 W, the growth rate increased with P_{rf}. Equation 1 presented below shows the growth rate (R) as a function of P_{rf} fitted by the Origin program

$$R = 0.0123 \times P_{rf, \, In2Ga2ZnO7} \qquad [1]$$

This increase indicates that the number of atoms sputtered from the target is proportional to the rf power. Comparisons of the experimental and calculated data are listed in Table 1. The calculated growth rate is 0.615 Å per sec at P_{rf} = 50 W. However, once P_{rf} was higher than 70 W, both the thickness uniformity and morphology of the deposited film became inferior. Furthermore, when we increased the substrate temperature from 25 to 275 ℃, there was no significant change in the growth rate.

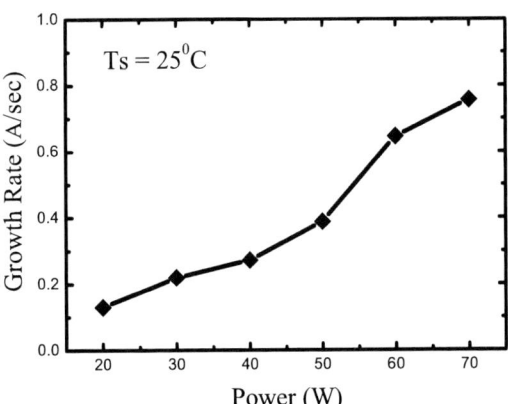

Figure 1. Growth rate of IGZO films as a function of rf power supplied to the $In_2Ga_2ZnO_7$ target (T_s = 25 ℃).

TABLE I. Comparisons of experimental and calculated growth rates with respect to different rf powers

$P_{rf, In_2Ga_2ZnO_7}$ (W)	Experimental Growth Rate (Å/sec)	Calculated Growth Rate (Å/sec)
20	0.130	0.246
30	0.220	0.369
40	0.272	0.492
50	0.388	0.615
60	0.646	0.738
70	0.756	0.861

3.1.2 Structural Properties

Figure 2 shows the X-ray diffraction patterns of IGZO films deposited on glass substrates. Film thickness was around 200 nm, and the substrate temperature during the sputtering process was 25 and 175 ℃ for samples (a) and (b), respectively. Two broad peaks were noted around 20° and 34° for both samples. The first was due to the glass substrate. The latter originated from the film and no sharp peak feature was found. This result indicates that deposited IGZO films are amorphous, which is consistent with the previous observation that the amorphous structure of the IGZO films is stable up to around 500℃ in air (20).

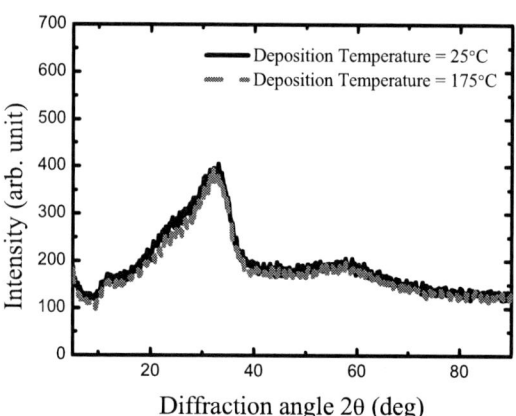

Figure 2. XRD patterns of IGZO films deposited on glass substrates at (a) 25 ℃ and (b) 175 ℃ by sputtering.

3.1.3 Electrical Properties

Figure 3 highlights temperature dependence of electrical resistivity for deposited IGZO films. The film thickness was 100 nm, for the four-point probe measurement, and substrate temperature during the sputtering was varied from 25 to 275 °C. No post-annealing was carried out on all the sample sets. The resistivity indicates a sharp decrease with increasing substrate temperature (25-175 °C), and less change at higher substrate temperatures. The minimum resistivity is 3.1×10^{-2} Ω-cm obtained at 175 °C, which is about ten times lower than that of the films deposited at room temperature. Clearly, electrical resistivity shows a thermally activated behavior.

It is a known fact that the film properties such as crystallinity, surface mobility during film growth, and the appearance of oxygen vacancies, are important factors which govern the electrical properties of oxide films. In order to analyze the conductive mechanism in IGZO films, we first investigate the XRD patterns. However, as shown in figure 1, no sharp peak is observed. The optimum deposition temperature of 175 °C, is far below the on-set crystalline temperature of about 500 °C for IGZO. Thus, the decrease in resistivity can be ascribed to two reasons. Firstly, with higher temperatures, atoms sputtered from target gain higher kinetic energy. As a result, atoms arrive at the film surface with higher surface mobility, which lead to higher film density and less dangling bonds. Additionally, it has been proven that the appearance of oxygen vacancies is a dominant factor for electrical conductivity of oxide semiconductors (23-25). Free electrons in TCO materials are mainly due to the generation of oxygen vacancies. Thermally excited oxygen atoms that can leave their original sites, induce vacancies with remaining free electrons at the sites. Excited oxygen atoms that have left their original sites will move into interstitial sites. The lower electrical resistivity observed at higher temperatures can be attributed to these free electrons generated alone with oxygen vacancies. Similar results have been found on indium tin oxide (ITO) compounds.

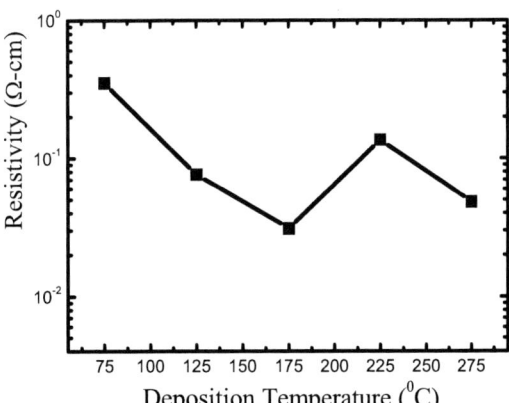

Figure 3. Electrical resistivity of IGZO films as a function of substrate temperature.

3.1.4 Optical Properties

Figure 4 shows the variation of optical transmittance (T) as a function of film thickness for IGZO films deposited at 175 °C. As the film thickness is 100 and 150 nm, the value of transmittance is over 99 %, around wavelength of 400 and 530 nm, respectively. It is concluded from Fig. 4 that as the film thickness increases, variations in transmittance are observed due to the interference phenomena. Figure 5 shows the relationship between optical band gap energy and the film thickness. The optical absorption coefficient, α, is defined as below

$$I = I_0 e^{-\alpha t} \qquad [2]$$

where I is the intensity of transmitted light, I_0 is the intensity of the incident light, and t is the thickness of the IGZO films. Since T is defined as I/I_0, we obtain α from equation [2]. And, optical band gap energy (E_g) and α are related by

$$\alpha = (\ hv - Eg\ )^{1/2} \qquad [3]$$

where h is the Planck's constant, and v is the frequency of the incident photon. From Fig. 5, the linear dependence of α^2 to hv indicates that IGZO films are direct-transition type semiconductors. The photon energy at the point where α^2 is zero is E_g. Thus, E_g can be determined using extrapolation methodology. It can be clearly seen that film thickness affects the band gap of deposited IGZO films.

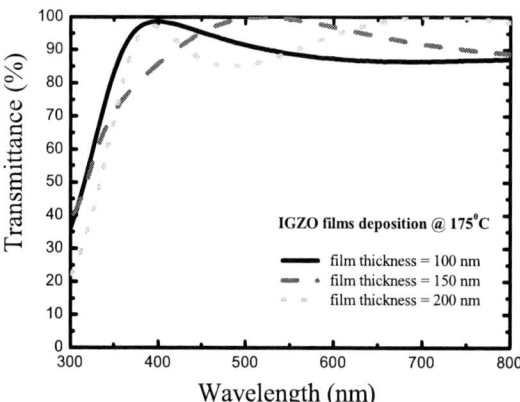

Figure 4. Effect of film thickness on optical transmittance spectra for IGZO films.

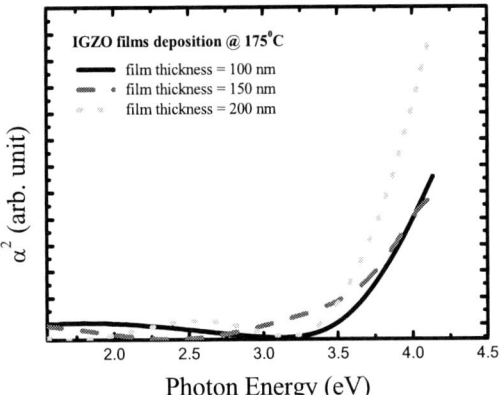

Figure 5. Square of the absorption coefficient as a function of photon energy for IGZO films with different film thicknesses.

3.2 IGZO Films Prepared by Co-sputtering $In_2Ga_2ZnO_7$ and In_2O_3 Targets

In this section, we deposited IGZO films by co-sputtering $In_2Ga_2ZnO_7$ and In_2O_3 targets. The rf power supplied to the $In_2Ga_2ZnO_7$ target was fixed at 50 W, while the rf power supplied to the In_2O_3 target was varied from 0 to 20 W. The substrate temperature was kept constant at 175 °C during the sputtering process. The structural, electrical, and optical properties of IGZO films are discussed in the following.

3.2.1 Structural Properties

Figure 6 shows the X-ray diffraction patterns for films deposited with different rf powers supplied to the In_2O_3 target. The film thickness was around 200 nm for all the samples. Besides the signal of the glass substrate being around 20°, we also observe two broad peaks around 34° and 58°. By increasing the In content in the IGZO films, the location of the measured diffraction peaks does not change significantly, however the intensity of the peaks becomes more pronounced. The reason for the formation of the halo peak around 34° may have contributed to the amorphous-like $Zn_kIn_2O_{3+k}$ structure (26, 27). However, it is still hard to confirm the exact crystalline phase with these broad peaks, and thus needs further research. Also observed in Fig. 6 is that only one strong peak is investigated around 34° for all the samples. This phenomenon can be understood by the survival of the fastest model proposed by Drift (28). According to this model, nucleation with various orientations can be formed at the initial stages of the deposition, and each nucleus competes to grow, but only the nuclei boasting the fastest growth rate can survive.

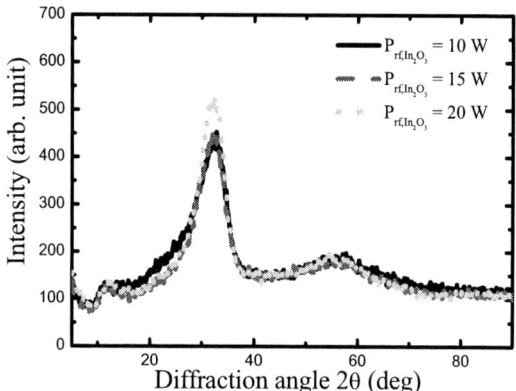

Figure 6. XRD patterns of IGZO films deposited on glass substrates with (a) $P_{rf, In2O3} = 10$ W, (b) $P_{rf, In2O3} = 15$ W, and (c) $P_{rf, In2O3} = 20$ W.

3.2.2 Electrical Properties

Figure 7 shows the variation of electrical resistivity, as a function of the rf power supplied to the In_2O_3 target. All the samples were deposited at 175 °C, and the film thickness was 100 nm. Electrical resistivity decreases with increasing rf power supplied to the In_2O_3 target. A minimum value of resistivity of 7.16×10^{-4} Ω-cm is obtained, while the rf power supplied to the In_2O_3 target is 20 W. Compared to IGZO films deposited by using a single $In_2Ga_2ZnO_7$ target, the minimum value of resistivity of IGZO films prepared by co-sputtering $In_2Ga_2ZnO_7$ and In_2O_3 targets, is lower than twice the order of magnitude. It has been generally known that the conduction characteristics of ZnO-based materials are primarily dominated by free electrons generated by oxygen vacancies and the Zn interstitial atoms [5, 6]. Thus, the decrease in resistivity is due to the contribution of In^{3+} and Ga^{3+} ions on substitutional sites of Zn^{2+} ions, which would result in a large number of zinc interstitials. In addition, the heavy metal In cations share electrons in 5s orbitals, and act as electron pathways, which is another important factor for the formation of more free electrons [7]. In order to clarify the relationship between resistivity and crystallinity, we compare Fig. 6 and 7. As the intensity of diffraction peaks increases, the resistivity decreases. This is an indication that resistivity is dependent on the crystallinity of the IGZO films. The improved crystallinity results in a lower density of traps for free carriers and the barriers for carrier transport in the film.

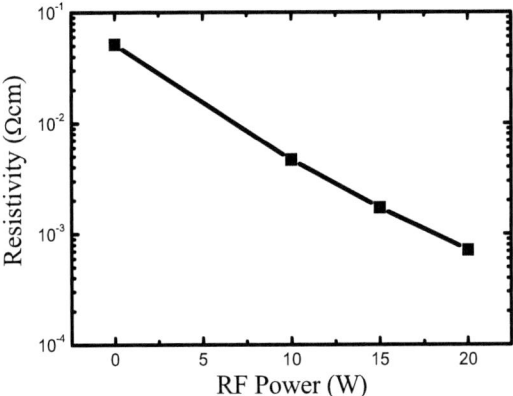

Figure 7. Electrical resistivity of IGZO films as a function of rf power supplied to the In_2O_3 target.

3.2.3 Optical Properties

Figure 8 shows the variation of optical transmittance as a function of the rf power supplied to the In_2O_3 target. Film thickness is measured at 100 nm, for all the samples. As rf power reaches 10 W, the value of transmittance is higher than 99 % in the wavelength region ranging from 516 to 535 nm, which indicates a great potential for the application in green LED chips. Figure 9 shows optical band gap energy, as related to the rf power supplied to the In_2O_3 target. The energy gap of In_2O_3 is ~ 3.5–4.0 eV. Thus, it can be clearly noted that the optical band gap energy of deposited IGZO films increases with increasing rf power supplied to the In_2O_3 target.

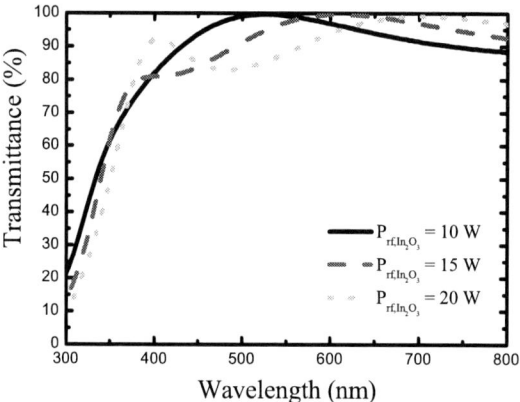

Figure 8. Optical transmission spectra of IGZO films deposited with (a) $P_{rf, In2O3}$ = 10 W, (b) $P_{rf, In2O3}$ = 15 W, and (c) $P_{rf, In2O3}$ = 20 W.

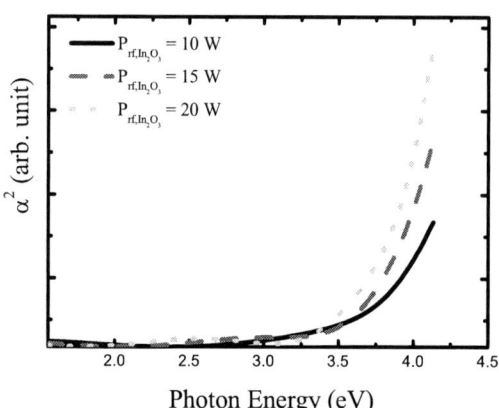

Figure 9. Square of the absorption coefficient as a function of photon energy for IGZO films with (a) $P_{rf, In2O3}$ = 10 W, (b) $P_{rf, In2O3}$ = 15 W, and (c) $P_{rf, In2O3}$ = 20 W.

4. Summary

In this paper, indium gallium zinc oxide (IGZO) thin films have been deposited by rf magnetron sputtering on glass substrates. The structural, electrical, and optical properties of these films are examined as functions of substrate temperature and the rf power supplied to the targets. In the first section, a single $In_2Ga_2ZnO_7$ target is used as a sputtering source. The growth rate increases with increasing rf power, which indicates that the number of atoms sputtered from the target is directly proportional to the rf power. The crystalline structures of films deposited at 25 and 175 °C are both amorphous, which is consistent with previous reports that state that the amorphous structure of IGZO films is stable up to around 500°C in air. The electrical resistivity decreases with increasing temperature. The optimum value of resistivity is 3.1×10^{-2} Ω-cm obtained at 175 °C. The main reason for the decreasing resistivity can be ascribed to less dangling bonds, and more thermally induced oxygen vacancies. As the film thickness increases, variations in optical transmittance are observed due to the interference phenomena. From the plot of α^2 to hv, it can be observed that IGZO films are direct-transition type semiconductors. Conjointly, the film thickness affects the optical band gap of deposited IGZO films.

In the second section, we prepared IGZO films by co-sputtering $In_2Ga_2ZnO_7$ and In_2O_3 targets. As the rf power supplied to In_2O_3 target increases, the halo peak around 34 ° becomes more distinct. This peak may be contributed to the amorphous-like $Zn_kIn_2O_{3+k}$ structure. The electrical resistivity decreases with increasing rf power supplied to the In_2O_3 target. A minimum value of resistivity of 7.16×10^{-4} Ω-cm is obtained while the rf power supplied to the In_2O_3 target was measured at 20 W. The decrease of resistivity is mainly contributed to two reasons. The first comes from In^{3+} and Ga^{3+} ions on substitutional sites of Zn^{2+} ions, and secondly, heavy metal In cations act as electron pathways. Both of them result in a large number of free carriers. The optical band gap energy of the deposited IGZO films increases while increasing the rf power supplied to the In_2O_3 target. As the rf power supplied to the In_2O_3 target equals to 10 W, the optical transmittance is higher than 99 % in the wavelength region ranging from 516 to 535 nm. These results indicate that the IGZO is the strongest candidate that can be used on GaN LED in the future.

References

1. H. K. Kim, K. S. Lee, and J. H. Kwon, *Appl. Phys. Lett.*, **88**, 012103 (2006).
2. K. Ramamoorthy, K. Kumar, R. Chandramohan, and K. Sankaranarayanan, *Mater. Sci. Eng. B*, **126**, 1 (2006).
3. R. H. Franken, C. H. M. van der Werf, J. Loffler, J. K. Rath, and R. E. I. Schropp, *Thin Solid Films*, **501**, 47 (2006).
4. J. O. Song, D. S. Leem, J. S. Kwak, Y. Park, S. W. Chae, and T. Y. Seong, *IEEE Photo. Technol. Lett.*, **17**, 291 (2005).
5. Y. R. Ryu, T. S. Lee, J. A. Lubguban, H. W. White, Y. S. Park, and C. J. Youn, *Appl. Phys. Lett.*, **87**, 153504 (2005).
6. M. Buchanan, J. B. Webb, and D. F. Williams, *Appl. Phys. Lett.*, **37**, 213 (1980).
7. T. Maruyama and K. Fukui, *Thin Solid Films*, **203**, 297 (1991).
8. S. Kulaszewicz, W. Jarmoc, and K. Turowska, *Thin Solid Films*, **112**, 313 (1984).

9. H. Kim, C. M. Gilmore, A. Pique, J. S. Horwitz, H. Mattoussi, H. Murata, Z. H. Kafafi, and D. B. Chrisey, *J. Appl. Phys.*, **86**, 6451 (1999).
10. Y. C. Lin, S. J. Chang, Y. K. Su, T. Y. Tsai, C. S. Chamg, S. C. Shei, C. W. Kuo, and S. C. Chen, *Solid-State Electron.*, **47**, 849 (2003).
11. E. M. C. Fortunato, P. M. Barquinha, A. C. M. B. G. Pimentel, A. M. F. Goncalves, A. J. S. Marques, R. F. P. Martins, and L. M. N. Pereira, *Appl. Phys. Lett.*, **85**, 2541 (2004).
12. N. L. Dehuff, E. S. Kettenring, D. Hong, H. Q. Chiang, J. F. Wager, R. L. Hoffman, C. H. Park, and D. A. Keszler, *J. Appl. Phys.*, **97**, 064505 (2005).
13. K. Nomura, H. Ohta, A. Takagi, T. Kamiya, M. Hiromich, and H. Hosono, *Nature (London)*, **432**, 488 (2004).
14. A. Takagi, K. Nomura, H. Yanagi, T. Kamiya, M. Hirano, and H. Hosono, *Thin Solid Films*, **486**, 38 (2005).
15. W. Lim, S. H. Kim, Y. L. Wang, J. W. Lee, D. P. Norton, S. J. Pearton, F. Ren, and I. I. Kravchenko, *J. Vac. Sci. Technol. B*, **26**, 3 (2008).
16. K. Nomura, T. Kamiya, H. Ohta, T. Uruga, M. Hirano, and H. Hosono, *Phys. Rev. B: Condens. Matter*, **75**, 035212 (2007).
17. H. Yabuta, M. Sano, K. Abe, T. Aiba, T. Den, H. Kumomi, K. Nomura, T. Kamiya, and H. Hosono, *Appl. Phys. Lett.*, **89**, 112123 (2006).
18. W. Lim, S. H. Kim, Y. L. Wang, J. W. Lee, D. P. Norton, S. J. Pearton, F. Ren, and I. I. Kravchenko, *J. Electrochem. Soc.*, **155**, 6 (2008).
19. H. Q. Chiang, B. R. McFarlane, D. Hong, R. E. Presley, and J. F. Wager, *J. Non-Cryst. Solids*, **354**, 2826 (2008).
20. K. Nomura, H. Ohta, A. Takagi, T. Kamiya, M. Hirano, and H. Hosono, *J. Jpn. Appl. Phys.*, **45**, 4303 (2006).
21. H. Hosono, K. Nomura, Y. Ogo, T. Uruga, and T. Kamiya, *J. Non-Cryst. Solids*, **354**, 2796 (2008).
22. G. H. Kim, B. D. Ahn, H. S. Shin, W. H. Jeong, H. J. Kim, and H. J. Kim, *Appl. Phys. Lett.*, **94**, 233501 (2009).
23. P. Kofstad, *J. Phys. Chem. Solids*, **23**, 1571 (1962).
24. P. Bonasewicz, W. Hirschwald, and G. Neumann, *Phys. Status Solids A*, **97**, 593 (1986).
25. V. Gavryushin, G. Raciukaitis, D. Juodzbails, A. Kazlauskas, and V. Kubertavicius, *J. Cryst. Growth*, **138**, 924 (1994).
26. D. S. Liu, C. H. Lin, F. C. Tsai, and C. C. Wu, *J. Vac. Sci. Technol., A* **24**, 694 (2006).
27. T. Minami, T. Yamamoto, Y. Toda, and T. Miyata, *Thin Solid Films*, **373**, 189 (2000).
28. A. Van der Drift, *Philips Res. Rep.*, **22**, 267 (1967)
29. Y. Igasaki and H. Saito, *J. Appl. Phys.*, **69**, 2190 (1991).
30. F.R. Blom, F. C. M. Van de Pol, G. Bauhuis, and Th. J. A. Popma, *Thin Solid Films*, **204**, 365 (1991).
31. H. Hosono, *J. Non-Cryst. Solid*, **352**, 851 (2006).

ECS Transactions, 28 (4) 149-153 (2010)
10.1149/1.3377111 ©The Electrochemical Society

InGaN-Based Light Emitting Diodes with an AlN Sacrificial Buffer Layer for Chemical Lift-Off Process

Chia-Feng Lin*, Jing-Jie Dai, and Ming-Shiou Lin

Department Department of Materials Science and Engineering, National Chung Hsing University, 250 Kuo Kuang Rd. Taichung 402, Taiwan

The InGaN-based light-emitting diodes (LEDs) grown on triangle-shaped patterned sapphire substrates were separated through a chemical lift-off process by laterally etching an AlN sacrificial layer at the GaN/sapphire substrate interface. The lateral etching rate of the AlN buffer layer was calculated at 10μm/min for the 100μm-width LED chip that was lifted off from the sapphire substrate. A triangular-shaped hole structure and a hexagonal-shaped air-void structure were observed on the lift-off GaN surface that was transferred from the patterned sapphire substrate. The chemical lift-off process was achieved by using an AlN buffer layer as a sacrificial layer in a hot potassium hydroxide solution.

Introduction

High efficiency InGaN-based light-emitting diodes (LEDs) have been intensively

Gallium nitride materials have attracted considerable interest in the development of optoelectronic devices like light-emitting diodes (LEDs) and laser diodes. However, bright blue LEDs require an increase in their internal and external quantum efficiencies. The lower external quantum efficiency of the InGaN-based LEDs is due to a larger refractive index difference between the GaN layer and the surrounding air ($\Delta n \sim 1.5$). Bottom patterned Al_2O_3 substrates[1], top p-type GaN:Mg rough surface processes[2,3], the formation of photonic crystal structures[4,5], periodic deflector embedded structures[6], and selective oxidization on the mesa sidewall[7] through a photoelectrochemical (PEC) wet oxidation process have been used to increase light-extraction efficiency in InGaN-based LEDs on Al_2O_3 substrates. Fujii et al.[8] reported that a laser lift-off technique followed by an anisotropic etching process to roughen the surface - an n-side-up GaN-based LED with a hexagonal "conelike" surface - has been fabricated to increase extraction efficiency. Stocker et al.[9] reported that crystallographic wet chemical etchings of the n-GaN have (0001), $\{10\bar{1}0\}$, $\{10\bar{1}1\}$, $\{10\bar{1}2\}$ and $\{10\bar{1}3\}$ stable planes. A chemical lift-off (CLO) technique has been realized by using a CrN layer,[10] a ZnO layer[11], and a Si-doped n-GaN layer[12] as the sacrificial layers. Ha et al.[13] reported that vertical light-emitting diodes (LEDs) were successfully fabricated throug a chemical lift-off process using a selectively etched CrN buffer layer.

In this paper, a description is given on how an InGaN-based LED epitaxial layer was lifted off a patterned sapphire substrate through the lateral etching of an AlN sacrificial layer at the GaN/Al_2O_3 interface using a hot potassium hydroxide solution (KOH, 80°C). The LED epitaxial layer grown on the triangle-shaped sapphire substrate had an air-void structure that provided a higher lateral etching rate on the AlN sacrificial layer through the chemical lift-off process. The surface morphologies and the optical properties of the lifted off InGaN epitaxial layers are analyzed in detail.

Experiments

InGaN-based LED structures were grown on a patterned c-face (0001) 2"-diameter patterned sapphire substrate by using a metalorganic chemical vapor deposition (MOCVD) system. The triangle-shaped patterns (5μm-width, 2μm-height, and 5μm-spacing) were fabricated on a sapphire substrate. The LED structure consisted of a 30nm-thick AlN buffer layer, a 6.5μm-thick n-type GaN layer, ten pairs of InGaN/GaN multiple quantum wells (MQWs) active layers, and a 0.2μm-thick magnesium-doped p-type GaN layer. The active layers consisted of a 30Å-thick InGaN-well layer and a 70Å-thick GaN-barrier layer for the InGaN/GaN MQW LED structure. A 100μm-width mesa pattern of an LED chip was prepared for this experiment. A 500nm-thick SiOx layer was deposited on the LED epitaxial layer as a hard mask for the inductively coupled plasma (ICP) dry etching process. The chip region was defined using an ICP etcher with Cl_2 gas that reached to the sapphire substrate and exposed the AlN buffer layer for the CLO process. Then, the LED wafer was immersed in a hot 2.2M KOH solution (KOH, 80°C) for a 5-minute lateral wet etching process that occurred at the sacrificial AlN buffer layer. A lateral wet etching process on the AlN buffer layer and a bottom-up N-face crystallographic etching process to form the pyramidal N-face GaN structure occurred during the CLO process. A similar crystallographic etching process that formed a continuous pyramidal LED structure was discussed in detail in our previous report.[14] The geometric morphology of these LED structures was observed through a scanning electron microscopy (SEM) and the optical properties of the samples were measured on micro-photoluminescence (μ-PL) spectra, with a 5μm-diameter laser spot size, using a 40mW, 325nm He–Cd laser as the excited source. The μ-PL spectra were characterized by an optical spectrum analyzer (Ando-6315A).

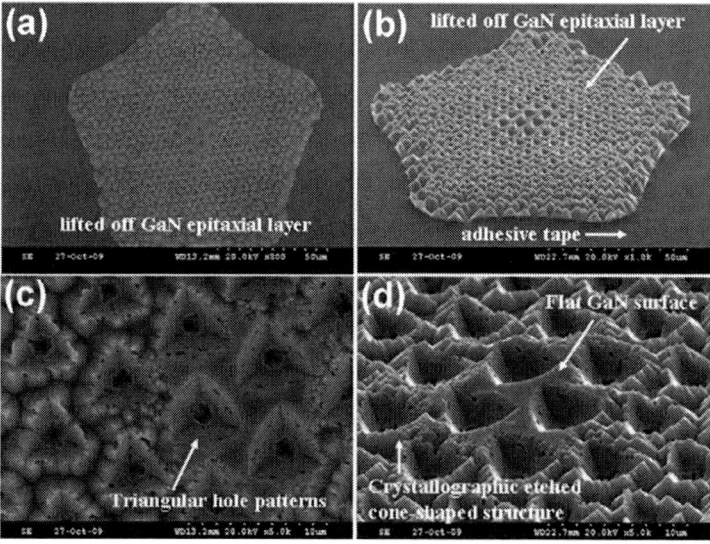

Figure 1. (a)(b) The SEM micrographs of the lift-off epitaxial layer are observed on the adhesive tape. The SEM micrographs, (c) top view and (d) 45-degree bird's eye view, of the lift-off epitaxial layer of the flat GaN surface at the chip center and of the cone-shaped GaN structure around the central region.

Results and Discussion

After the 5-min lateral wet etching process, the 100μm-width epitaxial layer was lifted off from the patterned Al_2O_3 substrate by using an adhesive tape. The lateral etching rate on the AlN buffer layer was calculated at 10μm/min. When the lateral etching width crosses the whole mesa region, the LED chip can be separated from the sapphire substrate. The SEM micrographs of the lift-off epitaxial film are observed in Fig. 1(a)(b). The shape of the truncated triangular pyramids on the sapphire substrate was transferred to the lift-off GaN film. The SEM micrographs of the bottom epitaxial layer are shown in Fig. 1(c) (top view) and Fig. 1(d) (45-degree bird's-eye view). In Fig. 1(c), the triangular patterns and the hexagonal shaped air-voids are observed on the lift-off GaN surface. At the center of the chip region, a partial flat GaN surface was formed by the laterally wet etched AlN buffer layer that separated it from the sapphire substrate. The cone-shaped structure of the N-face GaN surface was formed by a crystallographic wet etching process. Using a hot KOH solution should be considered in order to have a higher lateral etching rate on the AlN sacrificial layer and a lower crystallographic etching rate on the N-face GaN surface during the CLO process. A patterned sapphire structure, an AlN buffer layer, and an air-void structure were observed for the higher lateral wet etching rate during the CLO process. Small cone-shaped structures were observed around the center flat GaN surface as shown in Fig. 1(d).

When the LED chip was lifted off from the sapphire substrate by an adhesive tape, the bottom epitaxial layer was observed on the adhesive tape by an SEM micrograph as shown in Fig. 1(a). The lift-off epitaxial layer was transferred again by adhesive tape, so that the p-type GaN:Mg layer could be observed on the adhesive tape for the μ-PL measurement. The laser spots were focused on the center of the p-type GaN:Mg surface of the LED chip on the sapphire substrate and on the adhesive tape, and the μ-PL spectra were measured as shown in Fig. 2. The peak wavelength and line width of the μ-PL spectra were measured on the LED structure at the values of 454.7nm (22.1nm) for the sapphire substrate (mesa center region) and 451.8nm (25.2nm) for the adhesive tape (without sapphire substrate), respectively. After the transfer of the InGaN-based LED epitaxial layer from the sapphire substrate to the adhesive tape, the compressive strain between the GaN and the sapphire substrate was fully released. The adhesive tape did not add any stress to the lift-off LED epitaxial layer. The peak wavelength bluehift phenomenon of the lift-off LED epitaxial layer on the adhesive tape could be caused by a compressed strain reduction in the InGaN active layers. A blueshift phenomenon and a broadened line-width of the μ-PL spectrum were measured for the lift-off LED epitaxial layer similar to the one reported by Hsu et al.[15] The PL peak wavelengths are almost the same at 451.8nm for the laterally etched mesa region and the lift-off epitaxial layer, both without contact with the sapphire substrate. The PL intensity at the mesa edged region, with a 5μm-diameter laser spot size, was higher than at the lift-off epitaxial layer, because of the light reflected from the separated bottom sapphire substrate without the AlN buffer layer. Compared to the LED/sapphire structure without removing the AlN buffer layer, and after the CLO process, the InGaN active layer of the lift-off epitaxial layer had a PL wavelength blueshift phenomenon and a partially compressed strain release effect.

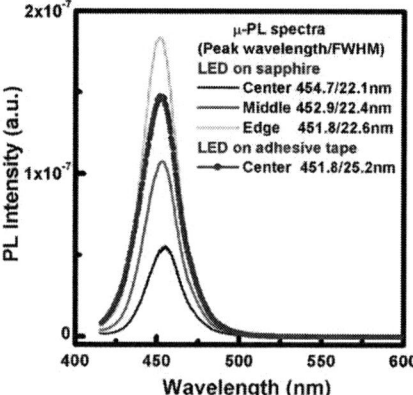

Figure 2. The μ-PL spectra of the partial laterally etched LED mesa structure are measured from the mesa center to the mesa edge at room temperature. The PL spectra of the lift-off epitaxial layer are also measured at the mesa center region.

Conclusions

In this letter, an InGaN-based LED structure grown on a sapphire substrate with an AlN buffer layer has been lifted off from the sapphire substrate through a CLO process. The AlN sacrificial layer and the air voids on the patterned sapphire substrate had a higher lateral wet etching process during the CLO process. The triangle-shaped-hole patterns, the hexagonal-shaped air voids, and the cone-shaped GaN structure were observed on the lifted off surface at the bottom of the LED structure. The free-standing lift-off InGaN-based epitaxial layer was realized through a CLO process and has the potential to replace the traditional laser lift-off process for vertical InGaN-based LED structure applications.

Acknowledgments

The authors gratefully acknowledge the financial support for this research from the National Science Council of Taiwan under grant No. NSC 98-2221-E-005-007-MY3 and No. NSC98-2622-E-005-009-CC3, and under the Ministry of Economic Affairs contract No. 97-EC-17-A-07-SI-097.

References

1. K. Tadatomo, H. Okagawa, Y. Ohuchi, T. Tsunekawa, Y. Imada, M. Kato, and T. Taguchi, Jpn. J. Appl. Phys., 40, p.583 (2001).

2. C. Huh, K. S. Lee, E. J. Kang, and S. J. Park, J. Appl. Phys. 93, p.9383 (2003).

3. S. J. Chang, L. W. Wu, Y. K. Su, Y. P. Hsu, W. C. Lai, J. M. Tsai, J. K. Sheu, and C. T. Lee, IEEE Photon. Technol. Lett., 16, p.1447 (2004).

4. A. David, T. Fujii, R. Sharma, K. Mcgroody, S. Nakamura, S. P. DenBaars, E. L. Hu, and C. Weisbuch, Appl. Phys. Lett. 88, 061124 (2006).

5. Hyun Kyong Cho, Junho Jang, Jeong-Hyeon Choi, Jaewan Choi, Jongwook Kim, Jeong Soo Lee, Beomseok Lee, Young Ho Choe, Ki-Dong Lee, Sang Hoon Kim, Kwyro Lee, Sun-Kyung Kim, and Yong-Hee Lee, Optics Express, Vol. 14, Issue 19, pp. 8654-8660 (2006)

6. H. G. Kim, M. G. Na, H. K. Kim, H. Y. Kim, J. H. Ryu, T. V. Cuong, and C.-H. Hong, Appl.Phys. Lett., vol. 90, p. 261117, 2007.

7. C. F Lin, Z. J. Yang, J. H. Zheng, and J. J. Dai, IEEE Photon. Technol. Lett., Vol. 17, No. 10, p.2038 (2005).

8. T. Fujii, Y. Gao, R. Sharma, E. L. Hu, S. P. DenBaars, and S. Nakamura, Appl. Phys. Lett. 84, 855 (2004).

9. D. A. Stocker, E. F. Schubert, and J. M. Redwing, Appl. Phys. Lett., 73, p.2654 (1998).

10. H. Goto, S. W. Lee, H. J. Lee, H.-J. Lee, J. S. Ha, M. W. Cho, and T. Yao,Phys. Status Solidi C 5, 1659 (2008).

11. D. J. Rogers, F. Hosseini Teherani, A. Ougazzaden, S. Gautier, L. Divay, A. Lusson, O. Durand, F. Wyczisk, G. Garry, T. Monteiro, M. R. Correira, M. Peres, A. Neves, D. McGrouther and J. N. Chapman, and M. Razeghi, , Appl Phys. Lett. 91, 071120 (2007)

12. Joonmo Park, Kwang Min Song, Seong-Ran Jeon, Jong Hyeob Baek, and Sang-Wan Ryu, "Doping selective lateral electrochemical etching of GaN for chemical lift-off", Appl Phys. Lett. 94, 221907 (2009)

13. J.-S. Ha, S. W. Lee, H.-J. Lee, H.-J. Lee, S. H. Lee, H. Goto, T. Kato, K. Fujii, M. W. Cho, and T. Yao, IEEE Photonics Technol. Lett. 20, 175 (2008).

14. Chia-Feng Lin, Chun-Min Lin, Chung-Chieh Yang, Wei-Kai Wang, Yu-Chieh Huang, Jien-An Chen, and Ray-Hua Horng, Electrochem. Solid-State Lett., Volume 12, Issue 7, pp. H233-H237, (2009)..

15. S. C. Hsu, B. J. Pong, W. H. Li, T. E Beechem, S. Graham and C. Y. Liu, "Stress relaxation in GaN by transfer bonding on Si substrates", Appl. Phys. Letts., vol. 91, (25), 251114 (2007)

ECS Transactions, 28 (4) 155-159 (2010)
10.1149/1.3377112 ©The Electrochemical Society

InGaN light-emitting diode structure on a photoelectrochemical treated GaN:Si layer

Kuei-Ting Chen, Chun-Min Lin, and Chia-Feng Lin

Department Department of Materials Science and Engineering, National Chung Hsing University, 250 Kuo Kuang Rd. Taichung 402, Taiwan

InGaN/GaN multiple-quantum-wells light-emitting diode (MQW-LED) structure was grown on a photoelectrochemical (PEC) treated n-type GaN:Si layer in deionized water. The PEC treated GaN:Si layer consisted of a nanoporous GaN:Si layer with a multiple-air-gaps (MAG) structure and the native GaOx layers that acted as submicro-masks for the epitaxial lateral overgrowth (ELOG) process. The threading dislocation density of the MAG-LED structure is seen reduced on the cross-sectional transmission electron microscopy micrographs. The photoluminescence spectrum has a 4.5nm wavelength blueshift phenomenon and a 2.0 times peak intensity enhancement in the MAG-LED structure (compared to a standard-LED) that is caused by a partially reduced piezoelectric field and an increasing light extraction process. The higher internal quantum efficiency and the higher light extraction process were observed in the MAG-LED structures by adding large volume of native GaOx layers and PEC treated multiple-air-gaps layers.

Introduction

High efficiency InGaN-based light-emitting diodes (LEDs) have been intensively investigated in recent year for various applications, such as traffic lights, back lighting in liquid crystal displays and solid-state lighting. However, the most common substrate for GaN epitaxial growth is sapphire (Al_2O_3) substrate that has a lattice mismatch of 14%. The lower internal quantum efficiency in InGaN layers is caused by the larger lattice mismatch inducing a higher dislocation density. Epitaxial growth technologies have been reported to reduce dislocation and increase light extraction efficiencies such as on patterned sapphires,[1] LEDs with an embedded photonic crystal,[2] and on a nanoporous GaN fabricated by ICP etching using anodic aluminum oxide (AAO) films acting as etch masks[3] for GaN epitaxial growth. We used the photoelectrochemical (PEC) oxidation process in deionized water to fabricate the nanostructure and native GaOx layers without using the wet etching process in acid solution and dry etching process. The GaOx layer is the native oxide layer in GaN material, and many articles have reported about the GaOx layers4 grown on the GaN surfaces through the PEC oxidation processes by using H_3PO[4,5] KOH,[6]or H_2O[7] solutions. The GaOx layers on the n-type GaN provided an enhancement of the photoluminescence (PL) and photocurrent (PC) responses due to a surface passivation effect, and a higher light extraction efficiency in InGaN-based LEDs due to a lower reflective index (n=1.89).[8]

In this paper, the nanoporous GaN:Si layer with the multiple-air-gaps (MAG) structure were fabricated through a photoelectrochemical (PEC) oxidation and oxide-

155

removing process in deionized water. This is the first time reports that the native GaOx layers acted as the self-assembled submicro-masks for the epitaxial lateral overgrowth (ELOG)[9] process and regrown the InGaN-based LED structures on the native GaOx layers. The threading dislocation density was reduced, the piezoelectric field in InGaN active layer was partially reduced, and the light extraction efficiency was increased in the InGaN-based LED structures by inserting the large volume of GaOx layer and the multiple-air-gaps structures. The optical and material properties of the InGaN-based LED structures grown on the PEC treated GaN:Si layers are discussed here in more detail.

Experiments

The InGaN-based LED structures were grown in a metal-organic chemical vapor deposition system on C-face (0001) 2" diameter sapphire substrates. The n-type GaN:Si layer consisted of a 30nm-thick GaN buffer layer grown at 550°C, a 1.2μm-thick unintentionally doped GaN layer, and a 2μm-thick n-type GaN:Si layer grown at 1150°C. The carrier concentration of the n-type GaN:Si layer was 1.5×10^{18} cm^{-3} analyzed by the Hall effect measurement. Two of the n-type GaN:Si epitaxial layer grown on sapphire substrates were prepared for the experiment. One of the wafer with a n-type GaN:Si layer was fabricated as a nanoporous structure through a photoelectrochemical (PEC) oxidation process in deionized water (DI) and an oxide-removed process in a diluted hydrochloric acid (HCl) solution. The wafer's edge for the external anode electrode was coated with the Indium (In) metal on the n-type GaN:Si layer. External bias was applied on the In metal on the n-type GaN:Si etching surface (the anode: In metal) and on the external Pt metal electrode (the cathode: Pt metal). The external bias was fixed at a positive 10V applied on the n-type GaN:Si layer during the PEC oxidation process, and the exposure time was 60 min with illumination from a 400W mercury (Hg) lamp. Using front-side illumination, the power density on the illuminated LED wafer was measured at around 86 mW/cm^2 (UV-C range: 220–290 nm). After removing the GaOx layer in the diluted HCl solution, the nanoporous structures were observed in the larger grain boundary regions on the n-type GaN:Si layer shown in Fig. 1(a). Then, a new GaOx layer was grown on the nanoporous GaN:Si layer through the same PEC oxidation process for 15 min to fill up the larger grain regions shown in Fig. 1(b). The PEC oxidation rate is about 250 nm/hr on the flat n-type GaN:Si layer. The other wafer with a n-type GaN:Si layer acted as the reference samples without PEC treatment for the following epi-growth process of the LED structure. The PEC treated wafer and the reference wafer were deposited the LED structures in MOCVD system at the same time. The regrown LED structures consisted of a 2μm-thick n-type GaN:Si layer (1150°C), eight pairs of In$_{0.2}$GaN/In$_{0.01}$GaN multiple-quantum wells (830°C), six periods of Al$_{0.3}$GaN/GaN (2nm/2nm) electron blocking layers (950°C), and a 0.13μm-thick p-type GaN:Mg layer (950°C). The active layers consisted of eight pairs each with a 3nm-thick In0.2GaN well layer and an 8.5nm-thick In$_{0.01}$GaN barrier layer. The surface morphologies and micro-structures were observed by using a field-emission scanning electron microscope (FE-SEM, JEOL 6700F) and a cross-sectional transmission electron microscopy (TEM). The power-dependent and temperature-dependent micro-photoluminescence (μ-PL) spectrums were measured by an optical spectrum analyzer using a 325 nm HeCd laser as the excited source.

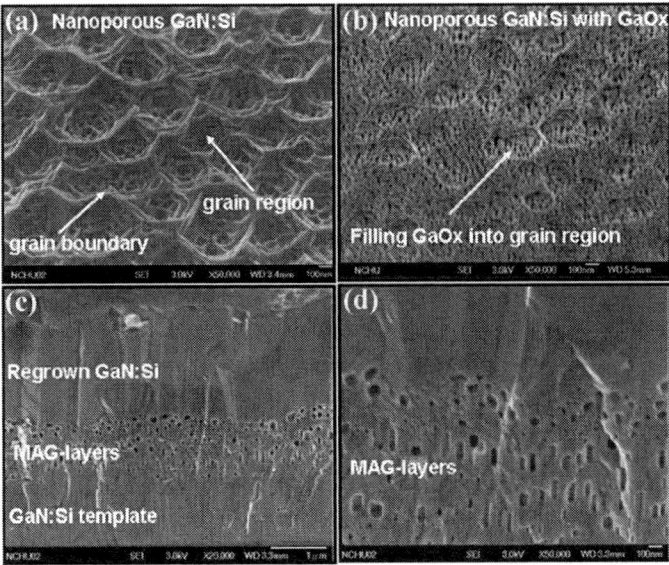

Figure 1. (a). SEM micrograph of a nanoporous GaN surface with smaller nanoporous structures and larger grain boundaries. (b). The GaO$_x$ layer grown in the larger grain region. (c)(d). Cross-section SEM micrographs of the MAG-LED structure with multiple air-gap layers.

Results and Discussion

After the PEC oxidation and oxide-removal processes, smaller nanoporous structures were observed in the larger grain boundary regions of the n-type GaN:Si template layer shown in Fig.1 (a). The dimensions of the average nanoporous and grain boundaries were measured at the values of 65-85nm and 420nm-460nm, respectively. Vajpeyi et al.[x] reported that the average pore size is highly sensitive to the dopant density of the as-grown GaN film prepared by UV-assisted electrochemical etching. In this experiment, the photo-induced electron-hole pairs were generated on the n-type GaN:Si layers, with a 1.5×10^{18} cm^{-3} carrier concentration, for the PEC oxidation process under Hg lamp illumination. After removing the original GaOx layer in the diluted HCl solution, the nanoporous GaN:Si structure with the bottom multiple-air-gaps (MAG) layers is formed in the GaN:Si epitaxial layer. Then, a new GaOx layer was grown on the nanoporous GaN:Si layer filling up the grain regions. The average nanoporous sizes of the GaOx layers were about 20nm-40nm as shown in Fig. 1(b). Initially, the GaOx layers grew in the grain regions and not on the grain boundaries on the nanoporous GaN:Si layer when the PEC oxidation rate at the grain boundaries was lower than at the grain regions.

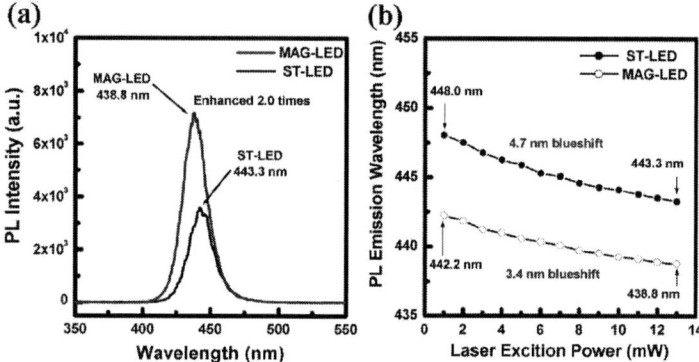

Figure 2. (a). The photoluminescence spectrums when analyzed by µ-PL measurement, and the peak wavelengths measured as 443.3nm for the ST-LED and 438.8nm for the MAG-LED structures. (b). Peak wavelengths of the PL spectra, measured at room temperature, for both LED samples when varying the laser excitation power.

From the µ-PL measurement in Fig. 2(a), the peak wavelengths of the PL emission spectrums were measured as 443.3nm and 438.8nm for the ST-LED and the MAG-LED structures, respectively. The peak emission intensity of the MAG-LED was enhanced 2.0 times compared to the ST-LED structure. The MAG-LED structure with the multiple-air-gaps layers and lower reflective index GaOx layer (n=1.89) can increase the light extraction efficiency to a higher PL intensity. The peak wavelengths of the PL spectra through varying the laser excitation power were measured at room temperature as shown in Fig. 2(b). The wavelength blueshift of the µ-PL spectrums were measured as 4.7nm for the ST-LED and 3.4nm for the MAG-LED structures by increasing the laser excitation power from 1mW to 13mW. The blueshift phenomenon was caused by the band filling effect when increasing the photo-induced carriers in the InGaN quantum well structure. In the power-dependent µ-PL measurement, the small wavelength blueshift phenomenon in the MAG-LED structure indicated a more flat band diagram in the InGaN well layer with a lower piezoelectric field after adding the GaOx layer and the multiple-air-gaps layers. During the PEC oxidation process, the GaOx layers formed above the multiple-air-gaps layers acted the mask layer during the ELOG process. The volume of a GaOx layer[xi] is larger than InGaN and GaN layer. The regrown n-type GaN:Si layer had the tensile effect that was caused by a large volume GaOx layer. The tensile effect on the n-type GaN:Si layer under the MQW active layer can partially release the compress strain in the InGaN well layers that the piezoelectric field was partially reduced.

Conclusions

In conclusion, the InGaN/GaN MQW LED structures were grown on a PEC treated n-type GaN:Si layer with multiple-air-gaps layers and a native GaOx layer. The GaOx layers grown in the grain regions acted as stop-growth submicro-masks for the epitaxial lateral overgrowth process in order to reduce the dislocation density. The blueshift phenomenon in the µ-PL spectrum and the smaller wavelength blueshift phenomenon in the power-dependent μ-PL measurement of the MAG-LED structures were caused by the partially compressed strain release in the InGaN well layer by inserting multiple-air-gaps layers and larger volume of GaOx layers in the LED structures. The lower threading dislocations density, lower piezoelectric field, and higher light extraction efficiency in the MAG-LED structures can be used for higher efficiency nitride-based LED applications.

Acknowledgments

This work was supported by the National Science Council under the Contact No. NSC 95-2221-E-005-132-MY3, NSC 97-2622-E-005-005-CC3, and NSC 97-2120-M-009-001.

References

1. D. S. Wuu, W. K. Wang, K. S. Wen, S. C. Huang, S. H. Lin, S. Y. Huang, C. F. Lin, and R. H. Horng, Appl. Phys. Lett. 89 (2006) 161105.
2. Min-Ki Kwon, Ja-Yeon Kim, Il-Kyu Park, Ki Seok Kim, Gun-Young Jung, Seong-Ju Park, Je Won Kim, and Yong Chun Kim, Appl. Phys. Lett. 92 (2008) 251110.
3. Y. D. Wang, K. Y. Zang, S. J. Chua, S. Tripathy, P. Chen, and C. G. Fonstad, Appl. Phys. Lett. 87 (2005) 251915.
4. L. H. Huang, S. H. Yeh, and C. T. Lee, Appl. Phys. Lett., vol. 93, 043511 (2008).
5. L. H. Peng, C. H. Liao, Y. C. Hsu, C. S. Jong, C. N. Huang, J. K. Ho, C. C. Chiu, and C. Y. Chen, Appl. Phys. Lett. 76 (2000) 511.
6. T. Rotter, D. Mistele, J. Stemmer, F. Fedler, J. Aderhold, J. Graul, V. Schwegler, C.Kirchner, M. Kamp, and M. Heuken, Appl. Phys. Lett. 76 (2000) 3923.
7. J. W. Seo, C. S. Oh, H. S. Jeong, J. W. Yang, K. Y. Lim, C. J. Yoon, and H. J. Lee, Appl. Phys. Lett. 81 (2002) 1029.
8. Chia-Feng Lin, Zhong-Jie Yang, Jing-Hui Zheng, and Jing-Jie Dai, J. Electrochem. Soc. 153 (2006) G39.
9. T. Mukai, K. Takekawa, and S. Nakamura, Jpn. J. Appl. Phys. 37 (1998) 839.
10. A. P. Vajpeyi, S. J. Chua, S. Tripathy, and E. A. Fitzgerald, Appl. Phys. Lett. 91 (2007) 083110.
11. V.M. Bermudez, "The structure of low-index surfaces of β-Ga2O3", Chemical Physics 323, 193–203 (2006).

160

Electron Paramagnetic Resonance Studies of Shallow Donors Behavior in Hydrogenated ZnO Films

L. L. Larina[a,c], N. A. Tsvetkov[c], J. H. Yang[b], K. S. Lim[b], and O. I. Shevaleevskiy[c]

[a] Department of Materials Science and Engineering, Korea Advanced Institute of Science and Technology, Yuseong, Daejeon 305-701, Republic of Korea
[b] Department of Electrical Engineering and Computer Science, Korea Advanced Institute of Science and Technology, Yuseong, Daejeon 305-701, Republic of Korea
[c] Solar Photovoltaic Laboratory, Semenov Institute of Chemical Physics RAS, Kosygin St. 4, 119334 Moscow, Russia

> Photoassisted metalorganic chemical vapor deposition (photo-MOCVD) technique was used for depositing of as-grown (ZnO) and hydrogenated (ZnO:H) zinc oxide polycrystalline thin films under different growth and hydrogen doping conditions. The structure and morphology of the films was controlled using SEM and AFM microscopy and surface roughness measurements. The studies of paramagnetic defects behavior and magnetic susceptibility in as-grown and hydrogenated samples were provided in the temperature range 77-300 K using X-band EPR spectrometer. It was found that all the samples exhibited a single symmetric strong Lorentzian line at g = 1.96 while the line intensity was shown to increase upon H-doping. The concentration of paramagnetic defects, associated with the concentration of the hydrogen donors, was extracted. The behavior of the magnetic susceptibility curves was described by Curie and Pauli contributions. We have shown that the excess concentration of paramagnetic defects, appeared after hydrogen doping, correlated with the increase of the Pauli term of magnetic susceptibility and followed the increase of free carrier concentration.

I. Introduction

In recent years different types of transparent conductive oxide (TCO) films were extensively investigated for solar photovoltaic devices application. Among them the layers of zinc oxide (ZnO) produced by different CVD technologies have gained an increasing interest due to successful application as p-type transparent wide band gap conductive front electrodes in amorphous and micromorphous silicon solar cells. The ability of low temperature ZnO films, deposition using low-pressure CVD process, make them promising for large-scale production of Si-based thin film solar modules [1]. Hydrogenated polycrystalline ZnO (ZnO:H) films are of special interest due to high electrical conductivity and stability of the photoelectronic parameters under solar illumination [2].

In our previous publications we reported on the improvement of electrical conductivity in the as-grown polycrystalline ZnO (poly-ZnO) films prepared by MOCVD via in situ hydrogen treatment [3-5]. We also provided EPR measurements of hydrogen-induced paramagnetic defects in poly-ZnO thin films that were associated with shallow donors [4]. In this study, we report on the preparation of hydrogen-doped polycrystalline

ZnO thin films (ZnO:H) using photo-MOCVD method. The deposited ZnO films were treated by in-situ post-deposition hydrogen doping using mercury-sensitized photodecomposition of hydrogen gas. The structure and morphology of the samples was controlled using SEM and AFM measurements. The behavior of hydrogen-induced paramagnetic native defects, associated with shallow donors, was investigated by Electron Paramagnetic Resonance (EPR) studies. The temperature dependent measurements of the defects behavior in ZnO:H films was studied by temperature-dependent measurements of magnetic susceptibility that provided us with the values of Pauli susceptibility contribution, associated with the excess concentration of the conduction electrons.

II. Experimental

Polycrystalline ZnO films were deposited on Corning 7059 glass substrates by photo-assisted metalorganic chemical vapor deposition (photo-MOCVD) technique. Bubbled diethylzinc (DEZ) and H_2O were introduced into the reaction chamber using a high purity Ar (99.999%) carrier gas. The H_2O/DEZ ratio and chamber pressure were maintained at 18 and 7 Torr, respectively. *In situ* post deposition hydrogen doping processes was performed at different temperatures of the samples ranging from 120 to 150°C. Hydrogen doping time was varied from 5 to 60 minutes. During the hydrogen doping process, the samples were irradiated by ultraviolet (UV) light perpendicularly to the substrate surface through a quartz window. A low-pressure Hg lamp with resonance lines of 184.9 and 253.7 nm was used as a UV light source. Before the introduction of H_2 gas into the reaction chamber, H_2 gas was passed through Hg bath in order to activate photodecomposition of H_2 (Hg-sensitized photodecomposition). The H_2 flow rate, chamber pressure, and Hg bath temperature were kept at 100 sccm, 1 Torr, and 20°C, respectively. A more detailed description of a deposition process can be found in our previous publications [5-7].

The structure and morphology of the samples was controlled using, SEM and AFM microscopy and roughness measurements. The studies of paramagnetic defects behaviour were provided using EPR X-band spectrometer Bruker EMX-8/2.7 operating under 100 kHz modulation frequency. To investigate the behaviour of the native defects in ZnO:H films temperature dependent measurements of EPR and of magnetic susceptibility in the temperature range 80-300 K were provided. The g-values were determined by measuring the difference between EPR resonance peak and that of a signal with a known g-value from Mn^{2+} containing probe. The absolute spin densities were calculated by comparing the doubly integrated values of EPR signal with a known paramagnetic standard.

III. Surface Morphology of ZnO:H Films

Figure 1 presents SEM images and shows the reorganization of the film surface in 4 poly-ZnO:H samples that were deposited using photo-MOCVD at 127, 135, 142 and 150°C and hydrogenated during 60 minutes. It is seen that the best quality of the films was gained for the samples deposited at 135 and 142°C.

The most critical parameter for TCO films that is important for the photovoltaic applications is the roughness of the sample surface. We have found that by varying the substrate temperature during the deposition process we can improve the roughness and find the optima substrate temperature. Figure 2 shows a comparative view of the AFM images for the 4 samples mentioned in a previous paragraph and deposited under the same experimental conditions.

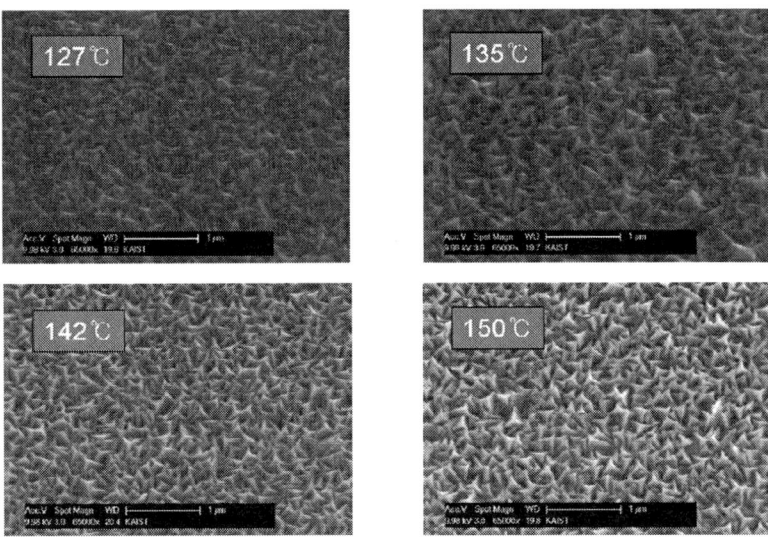

Figure 1. SEM images of poly-ZnO:H film deposited by photo-MOCVD under 127, 135, 142 and 150°C substrate temperature and hydrogenated during 60 minutes.

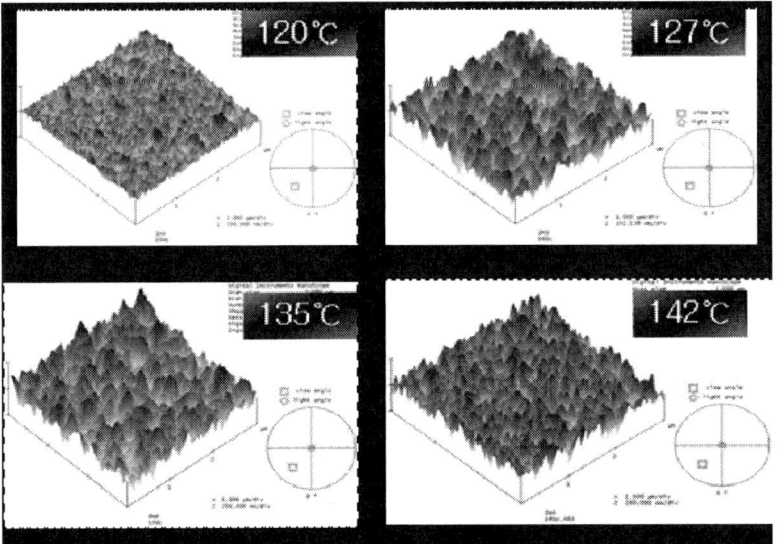

Figure 2. AFM images of-ZnO:H film surfaces prepared under 127, 135, 142 and 150°C substrate temperatures and hydrogenated during 60 minutes.

Figure 3 shows the dependence of ZnO:H layer roughness on the substrate temperature. This study provided us with the technological parameter of the optimum substrate temperature (135°C) for depositing the film with the increased roughness.

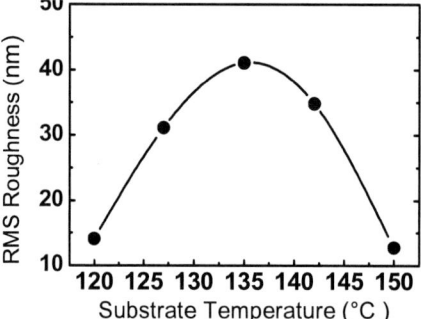

Figure 3. The roughness of ZnO:H film surfaces versus substrate temperature.

IV. EPR measurements of ZnO:H Films

Figure 4 depicts normalized EPR spectra recorded for poly-ZnO films before and after hydrogenation at 77, 210, and 300 K. All the EPR signals show the effective g-factor of 1.956 ± 0.001. The linewidth of ESR signal at 77 K, ΔH_{pp}, was found to be 0.63 ± 0.05 mT for as-grown poly-ZnO and 0.95 ± 0.05 mT for hydrogen treated poly-ZnO:H sample.

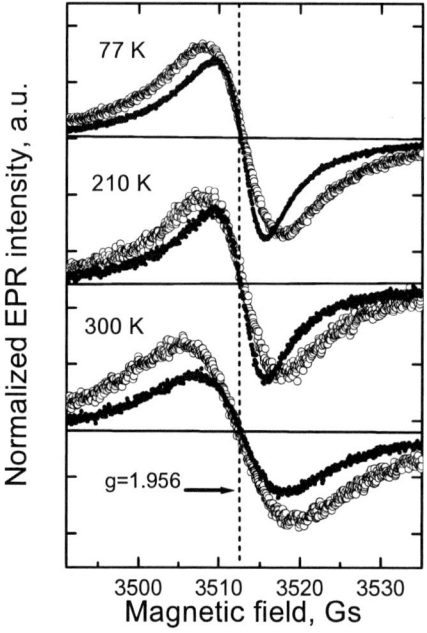

Figure 4. Normalized EPR spectra of shallow donors recorded at 77, 210, and 300 K for as-grown (dark circles) and hydrogenated (open circles) poly-ZnO films.

The EPR lines exhibit a Lorentzian line shape that can be attributed to either delocalized electrons in the conduction band or to a shallow impurity band of semiconductors. With increasing the temperature from 77 to 300 K the EPR signal for both types of poly-ZnO samples becomes less intense. The observed behavior can be explained by the decrease of the magnetic susceptibility of the paramagnetic centers in accordance with the Curie law. At the same time, in both cases the linewidth of EPR signals remain nearly constant with the temperature suggesting that the observed signals originated from the localized donor states. The concentration of paramagnetic centers was extracted by comparison to a reference standard. We have found that EPR signal for as-grown untreated poly-ZnO sample corresponds to 8.2×10^{17} cm^{-3} donors, while in hydrogen treated poly-ZnO:H thin films an appropriate value increased to 1.9×10^{18} cm^{-3}.

IV. Magnetic susceptibility of ZnO:H Films

Spin magnetic susceptibility $\chi_{spin}(T)$ as a function of temperature was calculated by comparing the integtated intensity of EPR signals of the samples and a reference paramagnetic standard. Fig. 5 shows the experimental curves of the normalized values of $\chi_{spin}(T)$ for poly-ZnO and poly-ZnO:H samples in the temparature range 77 – 300 K.

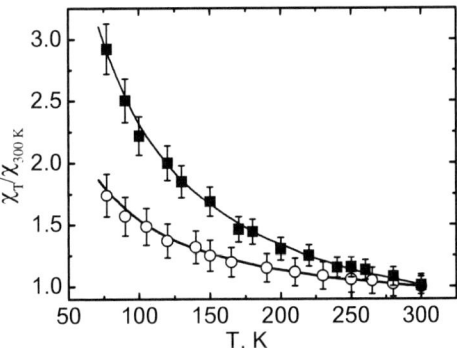

Figure 5. Temperature dependences of normalized spin magnetic susceptibility for poly-ZnO (circles) and poly-ZnO:H (quadrangles) samples.

Spin magnetic susceptibility of a paramagnetic sample consists of Curie and Pauli components:

$$\chi_{spin}(T) = \chi_{Curie}(T) + \chi_{Pauli} \tag{1}$$

χ_{Pauli} presents a temperature independent contribution initiated by the conducting electrons, $\chi_{Curie}(T)$ – is a temetrature dependent term that follows Curie law.
According to Curie law $\chi_{Cirue}(T)$ can be expressed as:

$$\chi_{Curie}(T) = C/T = \frac{N_S g^2 \mu_B^2 S(S+1)}{3 k_B \rho T}, \tag{2}$$

where N_S is volume concentration of paramagnetic centers, μ_B – Bohr magneton, k_B = Boltzmann constant, ρ - the density of the material, and C – Curie constant. Our studies have shown that spin concentrations N_S were found to be 8×10^{17} cm^{-3} and 2×10^{18} cm^{-3} appropriately for poly-ZnO and poly-ZnO:H samples.

Figure 6 shows the behavior of the magnetic susceptibility versus reversed temperature for the both types of ZnO samples under investigation.

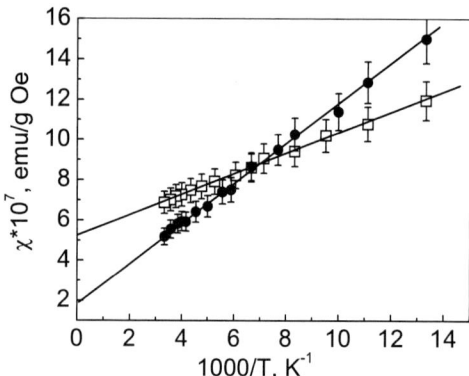

Figure 6. Spin magnetic susceptibility as a function of reversed temperature for poly-ZnO (circles) and poly-ZnO:H (quadrangles) samples.

Pauli susceptibility in semiconductor material is propotional to the density of the electronic states at the Fermi level ρ_0 (E_F) and can be expressed as:

$$\chi_{Pauli} = (g^2 \mu_B^2 / 2) \rho_0 (E_F) \qquad (3)$$

The values of χ_{Pauli} were extracted from the data presented in Fig. 6 by extrapolating the straight-line dependences of $\chi_{Spin}(1/T)$ to the zero value at the X-axis. It was found that after hydrogenation he appropriate values was 2.5 times increased from $\chi_{Pauli} = 0.2 \times 10^{-6}$ emu/g in poly-ZnO sample to 0.51×10^{-6} emu/g for poly-ZnO:H sample.

V. Shallow Donors Behavior in ZnO:H Films

We have shown that owing to EPR data the hydrogen doping of polycryatalline ZnO films iniciates the formation of hydrogen-related defects that can be associated with shallow donors. The density of the electronic states calcualted using Eq. (3) was found to increase from $\rho_0(E_F) = 0.9 \times 10^{21}$ eV^{-1}cm^{-3} in as-grown poly-ZnO to 2.24×10^{21} eV^{-1}cm^{-3} in hydrogenated poly-ZnO:H.

The electric conductivity measurements provided using 4-probe method have confirmed that hydrogenation process increases the conductivity of the samples that is in agreement with the observed increase of the density of the electronic states obtained from the magnetic susceptibility measurements.

The observed EPR lines with a Lorentzian line shape can be attributed to either delocalized electrons in the conduction band or to a shallow impurity band of semiconductors. As the temperature was increased from 77 to 300 K the ESR signal for both untreated and hydrogen-treated poly-ZnO samples became less intense. The observed behaviour correlates with the decrease of the magnetic susceptibility of the paramagnetic centres that follow the Curie law.

VI. Conclusions

To conclude, we used post-deposition hydrogen doping technique for producing p-type thin polycrystalline ZnO:H films for application in solar photovoltaics with the improved surface roughness, controlled amount of hydrogen donors and the improved electronic parameters. The concentration and the type of the shallow donors resulted from hydrogen doping was investigated by temperature dependent EPR and magnetic susceptibility measurements.

References

1. C. G. Van de Walle, *Phys. Rev. Lett.,* **85,** 1012 (2000).
2. R. G. Gordon, MRS Bull., **25,** 52 (2000).
3. S. Y. Myong, K. S. Lim, *Org. Electron.,* **8,** 51 (2007).
4. O. Shevaleevskiy, S. Y. Myong, N. Tsvetkov, and K. S. Lim, *J. Non-Cryst. Solids* **354,** 2849 (2008).
5. S.Y. Myong, K.S. Lim, Appl. Phys. Lett. **82,** 3026 (2003).
6. D. M. Hofmann, A. Hofstaetter, F. Leiter, H. Zhou, F. Henecker and B.K. Meyer, *Phys. Rev. Lett.,* **88,** 045504-1 (2002).
7. K. S. Lim, O. Shevaleevskiy, *Pure Appl. Chem.,* **80,** 2141 (2008).

168

ECS Transactions, 28 (4) 169-175 (2010)
10.1149/1.3377114 ©The Electrochemical Society

Dy^{3+} Emission from GaAlN Powder and Radio-Frequency Sputtered Thin Film

Jonathan H. Tao[a], Joanna McKittrick[a,b], Jan B. Talbot[a,c], and Kailash C. Mishra[d],

[a]Materials Science and Engineering Program, [b]Department of Mechanical and Aerospace Engineering, [c]Department of Nanoengineering,
University of California, San Diego, La Jolla, California 92093, USA
[d]Osram Sylvania, Central Research, Beverly, MA, 01915, USA

A three-step solution method previously used to synthesize AlN and GaN powders has been used to synthesize alloys of GaAlN. With the use of a stainless steel pressure vessel, single phase GaAlN powders were successfully synthesized. Powders of GaAlN activated with Dy^{3+} showed its related luminescence. The emission intensity of Dy^{3+} increased when Al was incorporated in the GaN host. Radio-frequency sputtering of the same source powder did not produce a film with the same composition as the source material; only an epitaxial GaN film was deposited. X-ray diffraction was used to characterize the phase of the powder and thin film. As Dy^{3+} emission was observed for GaAlN:Dy^{3+} but not for GaN:Dy^{3+} powders synthesized by the same approach. This suggests that Al incorporation allows the luminescence of this rare-earth activator in GaN.

Introduction

Rare-earth (RE) ions have been used as activators in oxide, phosphate, and vanadate compounds for decades as fluorescent lamp phosphors (1, 2). For this reason, there has been active research in incorporating RE ions into a nitride host for light-emitting devices (3). Eu^{3+}, Tb^{3+}, Dy^{3+}, and Tm^{3+} have demonstrated emission in the red, green, yellow and blue region of the visible spectrum in AlN powders, respectively, as we have shown in our past work (4, 5), as well as with Tb^{3+}-activated GaN powders and pulsed-laser deposited thin films (6). REs have also been used successfully in a GaN host as color emitters in electroluminescent thin films and devices (3, 7-9). Furthermore, the band gap of GaN (3.4 eV) can be tuned by alloying with AlN (6.2 eV), which allows the exploitation of higher RE energy levels (10); visible emission from implanted Eu^{3+}, Er^{3+}, Tb^{3+} and Tm^{3+} ions have also been shown to improve in GaAlN films grown by metal organic vapor phase epitaxy (MOVPE) compared to GaN (11-14).

GaAlN alloys are most commonly fabricated as thin films using MOVPE or sometimes by halide vapor phase epitaxy, while RE are ion implanted after the films are fabricated (10, 14-16). In the current work, we present a simpler alternative for synthesizing nitride alloys and incorporating RE ions. The three-step solution method used previously to synthesize AlN and GaN powders (17) has been used to synthesize alloys of $Ga_{1-x}Al_xN$ ($0.1 < x < 0.5$) powders. With the use of an additional stainless steel pressure vessel (Parr bomb), we have synthesized single phase $Ga_{1-x}Al_xN$ powders (x = 0.5). To study the effect of aluminum incorporation on Dy^{3+} luminescence, Dy^{3+} (1 atom %) has been incorporated into the nitride alloy at two different concentrations of Al, namely

169

$Ga_{1-x-y}Al_xDy_yN$ (x = 0.1 and 0.3; y = 0.01). Subsequently, $Ga_{0.69}Al_{0.30}Dy_{0.01}N$ powders were used as the source material to radio-frequency (RF) sputter thin films onto a sapphire substrate. X-ray diffraction (XRD) was used to characterize the phase of the powder and thin films. Cathodoluminescence (CL) was measured at room temperature for all powders and thin films, which were excited at 5 keV and 200 μA.

Experimental Procedures

Powder Synthesis

The process of preparing the $Ga_{1-x}Al_xN$ and $Ga_{1-x}Al_xDy_yN$ powders consisted of multiple steps. For each step, the amount of reactants was calculated and weighed based on the desired product weight and 100% conversion.

To prepare the $Ga_{1-x}Al_xN$ powder, first the desired amount of host nitrate, $Ga(NO_3)_3 \cdot XH_2O$ (Puratronic by Alfa-Aesar, 99.999%) and $Al(NO_3)_3 \cdot XH_2O$ (Puratronic by Alfa-Aesar, 99.999%), was dissolved in deionized (DI) water. The nitrate solution was then reacted with ammonium hydroxide (NH_4OH, Fisher Scientific, 28.50%) at room temperature until the hydroxide product, $Ga_{1-x}Al_x(OH)_3$ (x = 0.5), became a viscous gel. The gel was then filtered and dried in a dessicator and then the powder was ground. A small amount of DI water was added to the ground powder and placed in a Teflon-lined container inside a stainless steel vessel with screw top, as shown in Figure 1. The entire vessel is then heated at 100 °C for approximately 90 hours to promote the formation of $Ga_{1-x}Al_x(OH)_3$. The screw-top stainless steel vessel was required to maintain the pressurized environment inside the PTFE container. Following the heating process, the powder-like contents of the container was once again filtered and dried.

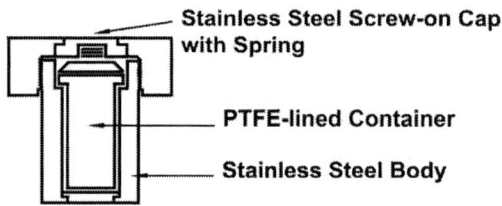

Figure 1. Schematic of the pressure vessel

To prepare the $Ga_{1-x-y}Al_xDy_yN$ powders, first Dy_2O_3 powder (REacton by Alfa Aesar, 99.9%) was dissolved in concentrated nitric acid (HNO_3, EM Science, 68.0 – 70.0%) that was heated to approximately 80°C, forming an aqueous solution of $Dy(NO_3)_3$. The activator solution was then added to the host nitrate solution of $Ga(NO_3)_3$ and $Al(NO_3)_3$ and placed in the stainless steel vessel to react as described above.

Next, the $Ga_{1-x}Al_x(OH)_3$ or $Ga_{1-x-y}Al_xDy_y(OH)_3$ was reacted with three times the stoichiometric amount of ammonium fluoride ($NH_4F \cdot XH_2O$, Alfa-Aesar, 99.9975%) in DI water and heated at approximately 80°C. After subsequent filtration and drying, the final step of the conversion process was introducing the hexafluoride $(NH_4)_3Ga_{1-x}Al_xF_6$ or $(NH_4)_3Ga_{1-x-y}Al_xDy_yF_6$ precipitate into a tube furnace to react with flowing ultrahigh-purity ammonia gas (Matheson Trigas, ULSI grade, 99.9995%) at 900°C for 150 min for the final conversion to the nitride alloy. This procedure, with the exception of the use of the pressure vessel, has been reported in detail elsewhere (17). For this study, the final compositions of the nitride powders were $Ga_{0.5}Al_{0.5}N$ for the undoped powder and $Ga_{0.89}Al_{0.10}Dy_{0.01}N$ and $Ga_{0.69}Al_{0.30}Dy_{0.01}N$ for the doped powders.

Radio-Frequency (RF) Sputtered Thin Film Deposition

The $Ga_{0.69}Al_{0.30}Dy_{0.01}N$ thin film was deposited from a source target of the same composition. Approximately 4 grams of $Ga_{0.69}Al_{0.30}Dy_{0.01}N$ powder was synthesized via the solution process with the use of the stainless steel vessel. Once the powders were thoroughly mixed, they were pressed into a pellet and annealed at 1100°C in NH_3. The pellet was then bonded via silver epoxy to a copper backing plate with an iron disk, held in place by a magnet at the end of the 33 mm RF sputtering gun (US Inc.). Sputtering was performed in N_2 at a pressure of 10^{-2} Torr in an ultrahigh vacuum chamber. The sputtering frequency was 13.56 MHz with a forward power of 100 W and no reflected power. The sputtering gun-substrate distance was 70 mm. The films were deposited on the c-plane of sapphire substrates, which were mounted onto a rotating heating stage, with a rotational speed of ~20 rpm. The deposition conditions were 500°C for 60 min. Following deposition, the film was annealed again at 900°C in NH_3 for 1 hr. It has been reported that annealing is necessary to activate the luminescent centers and improve the emission intensity (18). For this study, one film was deposited.

Results and Discussion

The XRD results in Figure 2 show the effect of using the stainless steel vessel when synthesizing the hydroxide precursor. The sample (a) synthesized without the stainless steel vessel separated into GaN and AlN phases, as indicated by the reference peaks of GaN and AlN. With the use of the stainless steel vessel, sample (b) showed a shift in the (100), (002), (101) peaks of GaN toward the corresponding (100), (002), (101) peaks of AlN, respectively, indicating the formation of an alloy. Annealing the nitride powder (sample (c)) for an additional hour at 1200°C further improved the alloy formation. This can most clearly be seen from peak (102), where the individual GaN and AlN peaks resolved into one broad peak. Lattice parameters were calculated by the software program Jade 5.0 (Materials Data, Inc.) from the peak position shift, as shown in Table I. The a and c lattice parameters both decreased, indicating the successful incorporation of Al into the GaN host. The DI water added to the $Ga_{0.5}Al_{0.5}(OH)_3$ acted as a transfer medium, allowing Al^{3+} ions to dissolve into the $Ga(OH)_3$ lattice when the hydroxide phase reached its solubility limit in the DI water. The stainless steel vessel heated at 100°C provided additional pressure to accelerate the dissolution process.

In addition to the $Ga_{0.5}Al_{0.5}N$ sample, Dy^{3+} doped GaAlN samples with 10 atom % Al were also synthesized with the stainless steel pressure vessel. The dominant transition of $^4F_{9/2} \rightarrow {}^6H_{13/2}$ for Dy^{3+} typically occurs around 575 nm. CL results shown in Figure 3 indicate a stronger Dy^{3+} emission in the $Ga_{0.89}Al_{0.10}Dy_{0.01}N$ powder compared to $Ga_{0.99}Dy_{0.01}N$ powder, where virtually no Dy^{3+} emission was observed. A very small and broad peak can be seen in Figure 3(b), the $Ga_{0.89}Al_{0.10}Dy_{0.01}N$ powder synthesized without the pressure vessel, but the peak became significantly stronger for the sample synthesized using the stainless steel vessel. This suggested that Dy^{3+} emission can be improved with the incorporation of Al into the GaN host, consistent with the observations by others (11, 12).

TABLE I. Lattice parameters of GaAlN powders synthesized with the pressure vessel.

Sample	Lattice parameter (Å)	
	a-axis	c-axis
GaN (00-050-0792)	3.189	5.186
AlN (00-025-1133)	3.111	4.979
(b) $Ga_{0.5}Al_{0.5}N$	3.162	5.179
(c) $Ga_{0.5}Al_{0.5}N$ annealed 1200°C	3.151	5.145

While the use of the pressure vessel aided the formation of the GaAlN alloy, the alloy could not be deposited as a film onto a sapphire substrate. Figure 4 shows the XRD data for the film deposited from $Ga_{0.69}Al_{0.30}Dy_{0.01}N$ source powder on a sapphire (006) substrate. Instead of a GaAlN film, an epitaxial GaN film was deposited, as the (002) peak position overlapped with that of the reference GaN. Although RF sputtering typically deposits a film with the same stoichiometry as the source material, the reason for this disparity between the compositions of the source material and the final film is still unclear. As seen in Figure 5, Dy^{3+} emission was still observed from the sputtered GaN film that was annealed at 900°C, but was weaker than the source powder. The decrease in Dy^{3+} emission may be attributed to the reduced Al content in the film compared to the source powder that seems to promote Dy^{3+} emission.

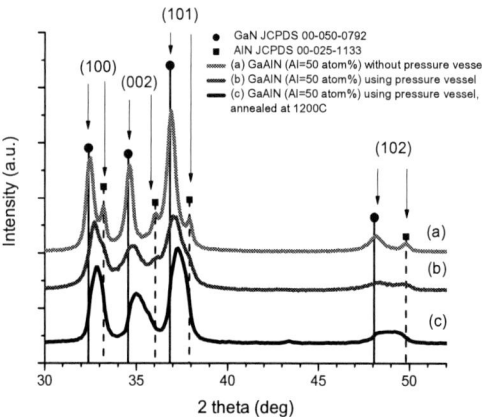

Figure 2. XRD of $Ga_{0.5}Al_{0.5}N$ powders synthesized from hydroxide precursor with and without using the pressure vessel, heated at 100°C.

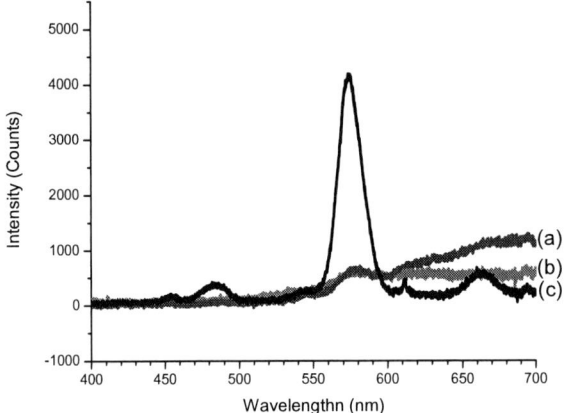

Figure 3. CL spectra of nitride alloy powders. (a) $Ga_{0.99}Dy_{0.01}N$ synthesized without pressure vessel, (b) $Ga_{0.89}Al_{0.10}Dy_{0.01}N$ synthesized without pressure vessel, (c) $Ga_{0.89}Al_{0.10}Dy_{0.01}N$ synthesized with pressure vessel.

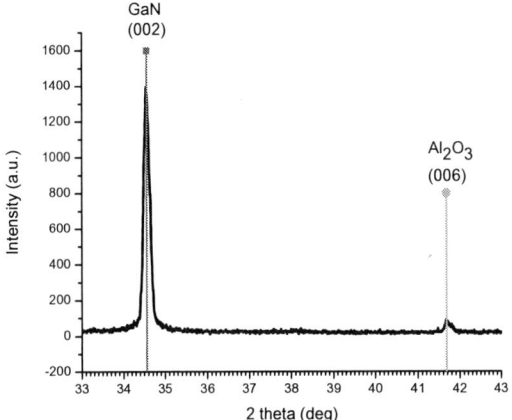

Figure 4. XRD of RF sputtered thin film from $Ga_{0.69}Al_{0.03}Dy_{0.01}N$ source target.

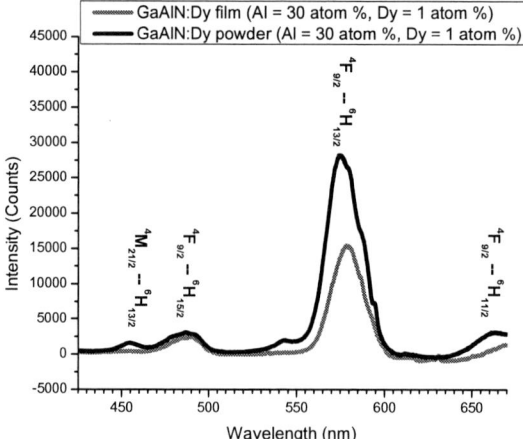

Figure 5. The various transitions for Dy^{3+} in $Ga_{0.69}Al_{0.30}Dy_{0.01}N$ powder and RF sputtered thin film. The transitions were assigned according to (4).

Conclusions

Using a pressure vessel, we were able to synthesize GaAlN powders that would have otherwise separated into GaN and AlN phases. Extended heating at 100°C in a pressurized environment was required to promote the formation of the alloy. Dy^{3+} dopant emission was also significantly enhanced in the GaAlN host compared to the GaN host. While the source powder was an alloy, the failure to deposit a GaAlN film could be due to the stoichiometry of the source material. The same method has been used to deposit pure GaN films made from commercial powder source that exhibited band gap emission at around 365 nm. Furthermore, XRD measurements indicate a GaN film was deposited with no band gap emission observed. This could be due to oxygen defects in the film that could not be identified by XRD, as the oxygen content was not enough to form a separate oxide phase. This is still under investigation.

Acknowledgments

The authors would like to thank Saul Peréz and Dr. Gustaf Arrhenius of Scripps Institute of Oceanography at UCSD for their insightful discussion and help with XRD measurements and other equipment. This project was supported by the Blasker-Rose-Miah Fund, grant no. C-2007-0024; and the Department of Energy grant DE-FC26-04NT42274, and DE-EE002003.

References

1. K. H. Butler, in *Fluorescent Lamp Phosphors*, p. 258, Pennsylvania State University Press, University Park (1980).

2. S. Kamiya and H. Mizuno, in *Phosphor Handbook*, 2 ed., W. M. Yen, S. Shionoya and H. Yamamoto Editors, CRC Press/Taylor and Francis, Boca Raton, FL (2006).
3. A. J. Steckl, J. C. Heikenfeld, D.-S. Lee, M. J. Garter, C. C. Baker, Y. Wang and R. Jones, *IEEE Journal of Selected Topics in Quantum Electronics*, **8**, 749 (2002).
4. B. Han, K. C. Mishra, M. Raukas, K. Klinedinst, J. Tao and J. B. Talbot, *J. Electrochem. Soc.*, **154**, J44 (2007).
5. B. Han, K. C. Mishra, M. Raukas, K. Klinedinst, J. Tao and J. B. Talbot, *J. Electrochem. Soc.*, **154**, J262 (2007).
6. J. H. Tao, J. Laski, N. Perea-Lopez, S. Shimizu, J. McKittrick, J. B. Talbot, K. C. Mishra, D. W. Hamby, M. Raukas, K. Klinedinst and G. Hirata, *J. Electrochem. Soc.*, **156**, J158 (2009).
7. Y. Q. Wang and A. J. Steckl, *Appl. Phys. Lett.*, **82**, 502 (2003).
8. J. M. Zavada, S. X. Jin, N. Nepal, J. Y. Lin, H. X. Jiang, P. Chow and B. Hertog, *Appl. Phys. Lett.*, **84**, 1061 (2004).
9. A. J. Steckl, J. Heikenfeld, D. S. Lee and M. Garter, *Mat. Sci. Eng. B*, **81**, 97 (2001).
10. K. Lorenz, E. Alves, T. Monteiro, A. Cruz and M. Peres, *Nucl. Intr. Met. in Phys. Res. B*, **257**, 307 (2007).
11. A. Wakahara, Y. Nakanishi, T. Fujiwara, H. Okada, A. Yoshida, T. Ohshima, T. Kamiya and Y. T. Kim, *Phys. Stat. Solidi A*, **201**, 2768 (2004).
12. A. Wakahara, Y. Nakanishi, T. Fujiwara, A. Yoshida, T. Ohshima and T. Kamiya, *Phys. Stat. Solidi A*, **202**, 863 (2005).
13. A. Wakahara, *Optical Materials*, **28**, 731 (2006).
14. I. S. Roqan, K. Lorenz, K. P. O'Donnell, C. Trager-Cowan, R. W. Martin, I. M. Watson and E. Alves, *Superlatt. Microstruc.*, **40**, 445 (2006).
15. J. Wu, W. Walukiewicz, K. M. Yu, J. W. A. III, S. X. Li, E. E. Haller, H. Lu and W. J. Schaff, *Sol. State Comm.*, **127**, 411 (2003).
16. K. E. Miyano, J. C. Woicik, L. H. Robins, C. E. Bouldin and D. K. Wickenden, *Appl. Phys. Lett.*, **70**, 2108 (1997).
17. J. H. Tao, N. Perea-Lopez, J. McKittrick, J. B. Talbot, B. Han, M. Raukas, K. Klinedinst and K. C. Mishra, *J. Electrochem. Soc.*, **155**, J137 (2008).
18. R. Weingärtner, O. Erlenbach, A. Winnacker, A. Welte, I. Brauer, H. Mendel, H. P. Strunk, C. T. M. Ribeiro and A. R. Zanatta, *Optical Materials*, **28**, 790 (2006).

CHAPTER 5

ENERGY DEVICES

178

ECS Transactions, 28 (4) 179-190 (2010)
10.1149/1.3377115 ©The Electrochemical Society

Cuprous Oxide Solution Preparation and Application to Cu₂O-ZnO Solar Cells

A. Du Pasquier[a], Z. Duan[b], N. Pereira[a] and Y. Lu[b]

[a] Department of Materials Science and Engineering, Rutgers University, North Brunswick NJ 08902
[b] Department of Electrical and Computer Engineering, Rutgers University, Piscataway, NJ 08854

Preparation of photoactive cuprous oxide (Cu_2O) is achieved via simple solution treatment of copper foils in copper sulfate aqueous solution. XRD indicates that pure Cu_2O phase is obtained, and gel electrolyte is used to probe high photocurrent on the film. Zinc oxide layer is deposited via MOCVD or spin coating, and the resulting Cu_2O-ZnO heterojunction solar cells are characterized with transparent Au or gallium doped zinc oxide (GZO) top contacts.

Introduction

Cuprous oxide (Cu_2O) is a promising low cost p-type photovoltaic absorber that can be grown by various methods including thermal oxidation of Cu foils (1) MOCVD (2) or electrochemical deposition (3). It is a natural p-type direct-gap semiconductor with bandgap energy of 2.1 eV. Calculations show that Cu_2O homojunctions have a theoretical energy conversion efficiency of 20% (4). Previously, it has been shown that one can deposit p-Cu_2O onto n-ZnO to realize a p–n heterojunction either by electrochemical deposition (5) or by magnetron sputtering (6). Several Transparent Conducting Oxide/Cu_2O heterojunctions have been tested, and good photovoltaic results ($\eta_{AM1.5}$=1.2%) were obtained with Al-doped ZnO-Cu_2O (7). The highest efficiency reported to date was $\eta_{AM1.5}$=2% in a sputtered MgF₂/ ITO/ZnO/Cu_2O device using MgF₂ as antireflection coating (8). However, Cu_2O external quantum efficiency is vastly dependant on its morphology, crystalinity and oxygen vacancy contents related to the preparation methods. Furthermore, many applications are forbidden by its instability. In this context, our motivation was to find a simple preparation method, and a practical measurement technique to evaluate the external quantum efficiency and the stability of Cu_2O samples for photovoltaic applications in combination with ZnO thin films.

Experimental

Cu₂O preparation

Copper foils (0.8 mm thick, 1*1 in² pieces) were mechanically micro polished, rinsed with methanol and treated for 10 min in a boiling aqueous solution of 10 wt.% $CuSO_4$. The foils rapidly turned to a purple color, indicating a conversion of Cu to Cu_2O according to the reaction:

$$Cu^0 + Cu^{II}SO_4 + H2O \rightarrow Cu^I_2O + H_2SO_4 \qquad [1]$$

The thickness of Cu_2O layer as function of reaction time was measured with a Veeco D150 stylus profilometer (Fig. 1). A maximum thickness of ~600 nm was reached after 10 minutes, and longer reaction times did not lead to thicker films. This is consistent with a chemical passivation process, where the reaction stops when all metallic copper has reacted on the surface.

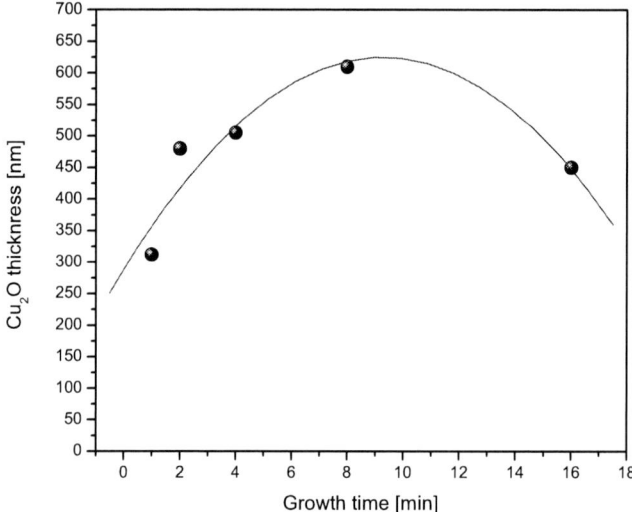

Figure 1. Profilometer Cu_2O thickness versus reaction time in boiling 10 wt.% $CuSO_4$ aqueous solution.

For comparison, we attempted the preparation of Cu_2O by oxidation of Cu foils in air at 700, 800 and 900°C. As expected, a mixture of CuO and Cu_2O phases was obtained, while the solution reaction led to pure Cu_2O phase, as evidenced by their XRD patterns (Fig. 2). Low oxygen contents are necessary to successfully prepare pure Cu_2O via thermal oxidation. Thus, the solution reaction presented here is advantageous for our purposes because it yields pure Cu_2O films, and requires considerably less energy than the thermal oxidation.

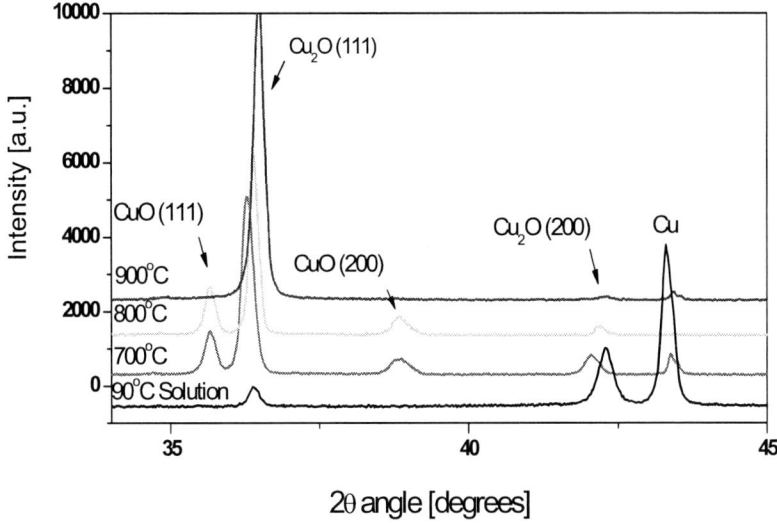

Figure 2. X-ray diffractograms of the Cu_2O phases grown for 10 min in air at 700, 800 and 900°C, and in solution at 90°C.

Cu₂O Photocurrent measurement with gel electrolyte.

Photocurrent measurement was achieved very simply with a fiber optics probe terminated by a copper contact, and using a Teflon spacer and gel electrolyte directly applied to the surface of Cu_2O. Monochromatic chopped light (40Hz) was directed to the sample through optical fibers, form an Oriel Cornerstone 130 monochromator illuminated with a Xenon light source. Upon excitation of the bandgap, holes or electrons react with the gel electrolyte, causing a photocurrent, which is collected by the metal surrounding the probe. This small AC photocurrent is amplified by lock-in amplifier (EG&G 5210), and the photocurrent spectrum is obtained. Calibration of the light input power with a photodiode enables to calculate the external quantum efficiency of the film. The Teflon spacer fixes the distance from probe to substrate, and the device area is fixed to 0.03 cm² by the illumination spot diameter (Figure 3). This method was very effective to characterize the photoconductivity of as prepared Cu_2O films. An aqueous gel for photocurrent measurement was prepared as follows: 2.7 g lactic acid and 1g $CuSO_4$ were dissolved in 10 mL deionized water and gelled with 0.5g Methocel. This resulted in a gel of 3 mol/L lactic acid and 0.4 mol/L $CuSO_4$. The photocurrent spectra were measured with a fiber optics reflection probe (Ocean Optics) terminated by the photocurrent measurement electrode. A small amount of gel was applied to the surface of the Cu_2O film, and the probe was applied on the gel. Electrical connections were made between probe and sample to the lock-in amplifier.

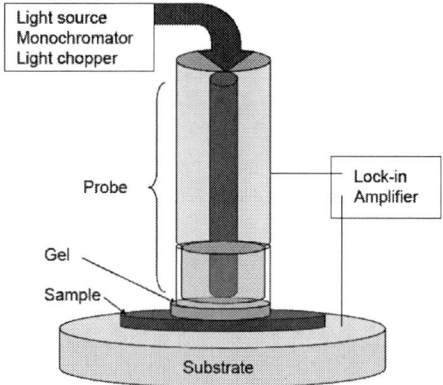

Figure 3. The photocurrent probe and the configuration used to measure photocurrent on a Cu_2O foil.

Preparation of photovoltaic devices

Several types of devices have been made using Cu_2O substrates prepared by the solution process described above; They were of the types $Cu_2O/ZnO/GZO$, where ZnO layers were deposited using two methods: (i) spin coating using 10 wt.% Zinc acetate in methanol, followed by 350°C annealing (9), or MOCVD using diethyl zinc (10). The transparent top contacts were either semi-transparent Au (7.5 nm) deposited by sputtering, or Ga-doped zinc oxide (GZO, 500nm), deposited by MOCVD (11). In the case of ZnO deposited by MOCVD, there is a preferential orientation along the c-axis, and dense arrays of essentially single crystal ZnO nanotips are formed (12). The morphology of the GZO layer is polycrystalline with no preferential orientation.

Photovoltaic devices were characterized by external quantum efficiency (EQE) using setup similar as used for photocurrent with gel electrolyte, and I-V in the dark and under AM 1.5 simulated sunlight.

Results and discussion

Gel measurements on Cu_2O films

Photocurrent versus wavelength where measured on Cu_2O films using the gel electrolyte and probe described in experimental section.

The photocurrents of Cu_2O films prepared either by oxidation at 900°C in air, or by solution reaction at 90°C in aqueous $CuSO_4$ are shown on Figure 4. The results indicate that higher photocurrents are obtained with the solution conversion process, which we attribute to the higher purity of the Cu_2O phase (absence of CuO evidenced by XRD).

We also observed that photocurrent stability versus time was greatly improved when using a non-aqueous gel electrolyte (Figure 5). The non-aqueous gel was prepared by dissolving 1g $CuSO_4$ in 10 mL N-methylpyrrolidinone (NMP), and gelling it with 1 g PVDF-HFP (Kynar 2801). We attribute the increased stability with the non-aqueous gel to the suppression of Cu_2O dissolution, which can proceed in aqueous $CuSO_4$ solution via the following reactions:

At $Cu/Cu_2O/CuO$ electrode:

$$\text{Hole formation in } Cu_2O: 2Cu_2O + h\nu \rightarrow 2Cu\square_2O + 2e\text{-} \qquad [2]$$

$$\text{Water oxidation: } H_2O \rightarrow 2H^+ + 1/2O_2 + 2e^- \qquad [3]$$

$$Cu_2O \text{ dissolution: } Cu_2O + 2H^+ \rightarrow 2Cu^{2+} H_2O \qquad [4]$$

At Cu electrode:

$$Cu^{2+} + 2e^- \rightarrow Cu \qquad [5]$$

This mechanism was confirmed by the visual evidence of Cu_2O dissolution after photocurrent measurement with the aqueous electrolyte, and the fact that dissolution only occurred when the circuit was closed and exposed to light. The photocurrent decay after ~3h with the non-aqueous gel is linked with solvent evaporation from the gel, since it was not hermetically sealed.

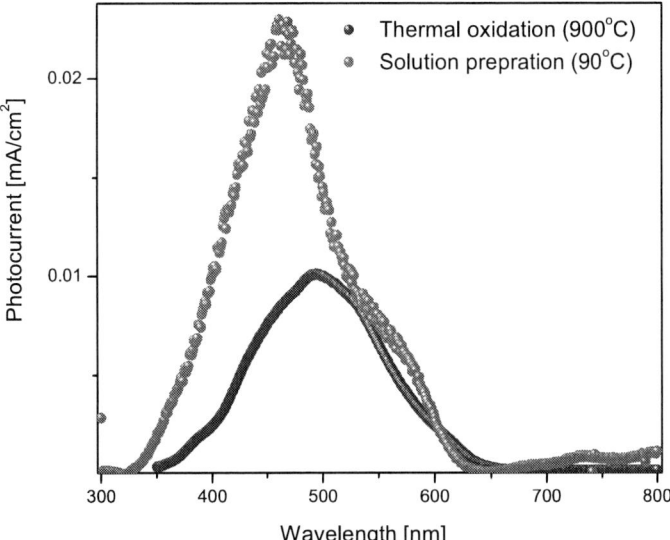

Figure 4. Photocurrent versus wavelength for Cu2O films prepared via thermal oxidation (900°C) or solution conversion (90°C) of copper foils.

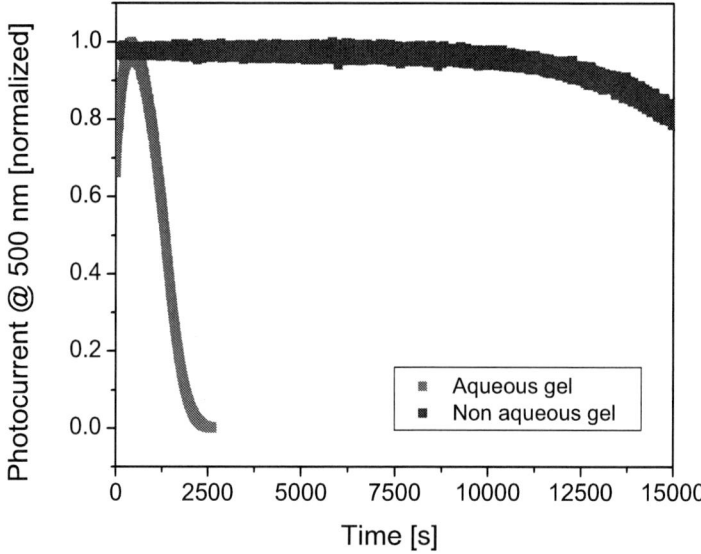

Figure 5. Photocurrent at 500 nm versus time for solution prepared Cu_2O films measured with aqueous or non-aqueous gel electrolytes

Photocurrent and EQE measurement of Cu_2O/ZnO photovoltaic cells

Effect of ZnO buffer layer. We compared the external quantum efficiency of $Cu/Cu_2O/ZnO_{50nm}/GZO_{500nm}$ and $Cu/Cu_2O /GZO_{500nm}$ where Cu/Cu_2O substrates are prepared by the solution conversion and coated with ZnO and GZO grown by MOCVD. The EQE measurements clearly show the beneficial effect of ZnO buffer layer on photocurrent. This is attributed to the formation of an n-p heterojunction which facilitates the separation of electron-hole pairs. An EQE of 9.5% was obtained with a 50 nm ZnO buffer layer (Fig. 6).

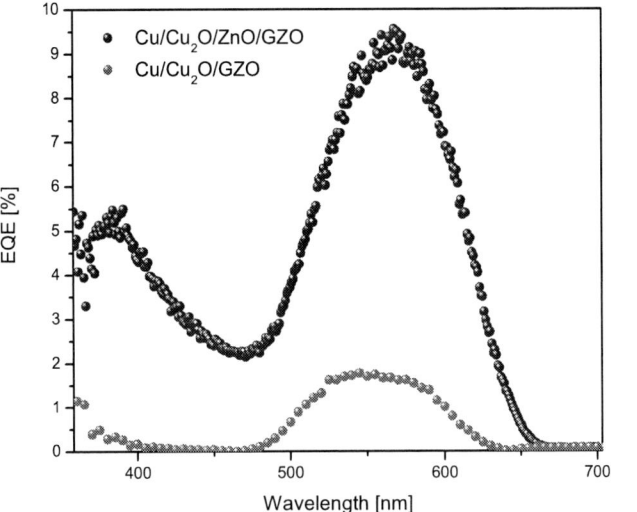

Figure 6. External quantum efficieny (EQE) of same $Cu/Cu_2O/GZO$ photovoltaic cells with or without 50nm MOCVD grown ZnO buffer layer between Cu_2O and GZO.

<u>Effect of voltage bias on photocurrent of Cu_2O/ZnO photovoltaic cells.</u> We studied the dependency of photocurrent versus wavelength at various voltage biases on $Cu/Cu_2O/ZnO/GZO$ devices. This is performed automatically with a Labview program which sets the bias on potentiostat and scans the wavelength on monochromator, while recording photocurrent with the Lock-in amplifier. The typical results shown on a 3D plot (Fig 7) indicate that voltage bias, either positive or negative, is necessary to obtain maximum photocurrent form the devices. The point of zero photocurrent corresponds to the open-circuit voltage, and was very low (<100 mV) in all devices tested. We conclude that such curves indicate a poor diode rectification, and low solar power conversion efficiency despite high photocurrent.

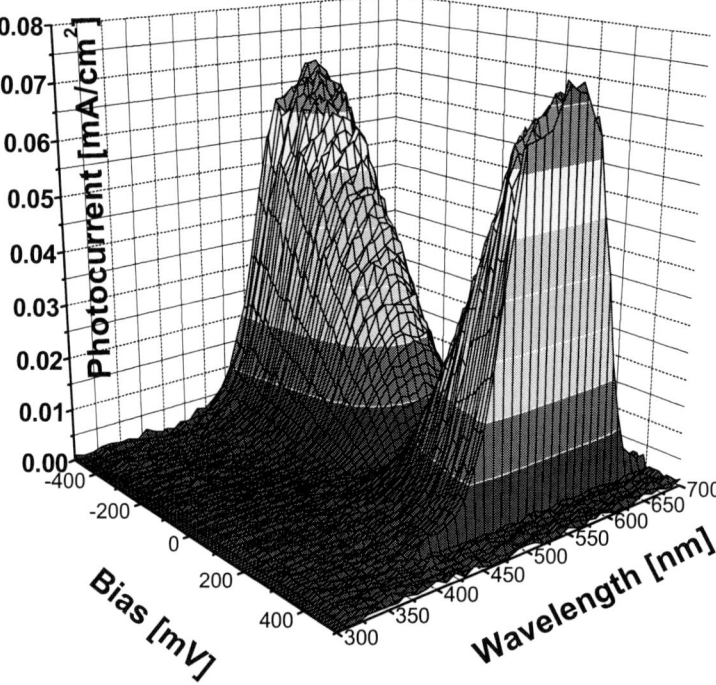

Figure 7. 3D plot of photocurrent vs. wavelength vs. bias for Cu/Cu$_2$O/ZnO/GZO

Effect of metal contacts on I-V response of Cu$_2$O/ZnO photovoltaic cells. We studied three configurations of Cu$_2$O/ZnO photovoltaic cells:

1. Cu/Cu$_2$O/ZnO/Au device, where the ZnO layer was spin-coated, and Au top contact was sputtered on ZnO. Such devices had almost no diode rectification, and no visible changes of the I-V curves during illumination. We attribute the poor diode rectification to the high work function of Au (5.1 eV), which opposes the built-in electric field at the heterojunction. (Fig. 8)

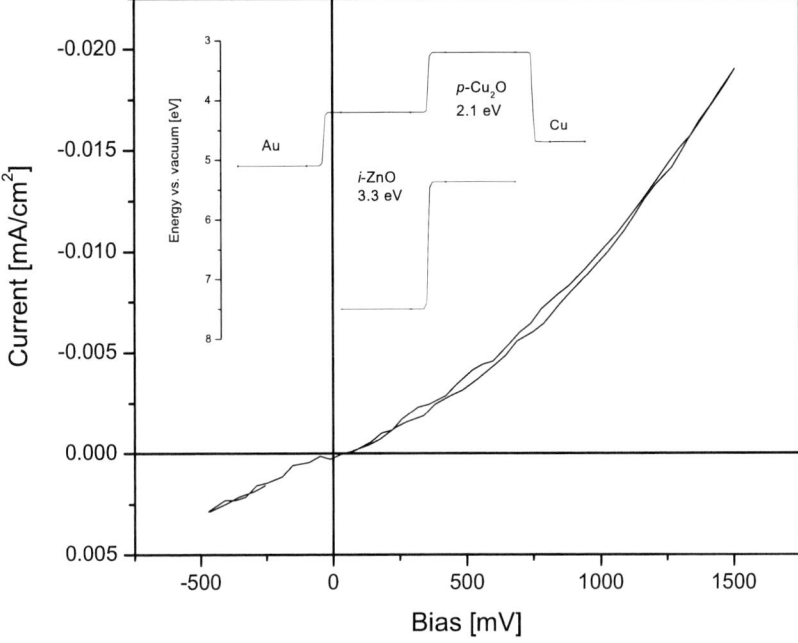

Figure 8. I-V plot in the dark for Cu/Cu$_2$O/ZnO/Au device (0.07 cm^2), with corresponding band diagram (inset)

2. Cu/Cu$_2$O/ZnO/GZO devices where the ZnO and GZO layers are grown by MOCVD (50 and 500 nm respectively). In this case, the diode rectification is improved, but photo response is still difficult to observe on the I-V curve. We attribute the improved diode rectification to the close conduction band alignment between GZO and ZnO. (Fig. 9)

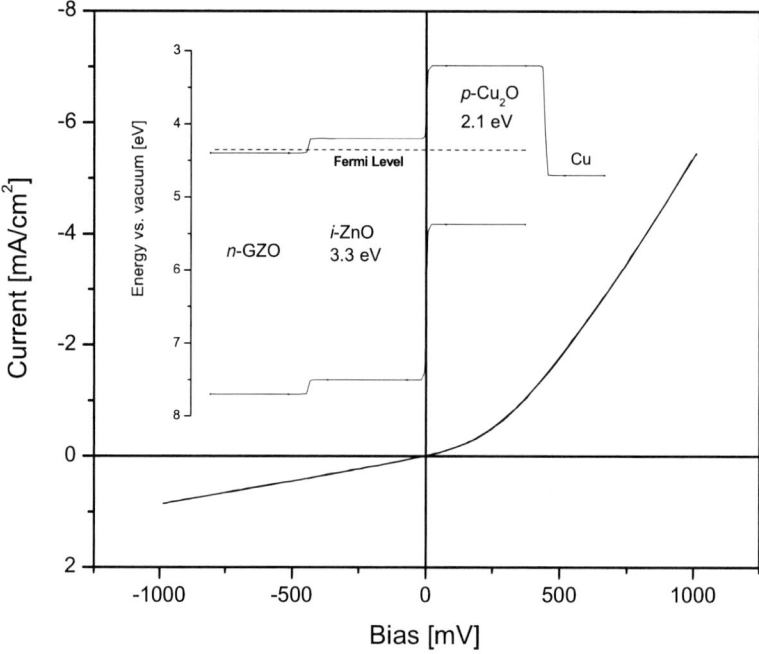

Figure 9. I-V plot in the dark and light for Cu/Cu$_2$O/ZnO/GZO device (0.07 cm^2) , with corresponding band diagram (inset)

3. FTO/ZnOnt/Cu$_2$O/Au , where ZnO was grown by MOCVD on commercial FTO/glass (Pilkington TEC15). Copper was electrodeposited on ZnO from an aqueous CuSO$_4$ solution. Then, copper film treated in the same boiling solution converts to Cu$_2$O according to the process previously described. Although Cu$_2$O can be directly electrodeposited, this is usually done at basic pH. We preferred this two step method to avoid dissolving the ZnO layer in basic solution. Then, 7.5 nm transparent gold contacts were sputtered on the Cu$_2$O surface. The devices obtained by this process had best diode rectification, and visible photoresponse under AM 1.5 illumination (Fig. 10). We attribute this result to the close band alignment between Cu$_2$O valence band and Au work function, which favors hole extraction from Cu$_2$O without opposing the electric field in the heterojunction.

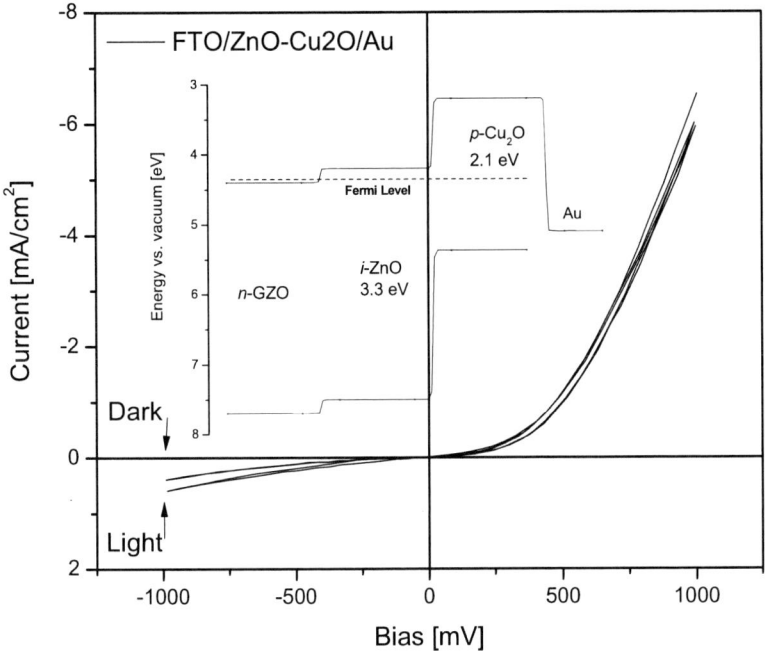

Figure 10. I-V plot in the dark and light for FTO/ZnOnt/Cu$_2$O/Au devices (0.07 cm^2), with corresponding band diagram (inset)

Conclusions

We have demonstrated a simple solution process for the preparation of pure phase Cu$_2$O, and shown higher photocurrents than with samples prepared by thermal oxidation. This process is considerably less energy consuming than the thermal oxidation, thus very well suited for photovoltaic production because it reduces energy payback time and lowers greenhouse gases emissions during production. We used a gel electrolyte and a simple probe for photocurrent measurement directly on the Cu$_2$O surface. We report much improved photocurrent stability with a non-aqueous gel electrolyte, and attribute photocurrent decay to Cu$_2$O dissolution with aqueous gel electrolyte. We used the solution prepared Cu$_2$O to build solid state Cu$_2$O/ZnO photovoltaic devices which demonstrate EQE of 9.5%. We show that best diode rectification is observed in devices of configuration GZO/ZnO/Cu$_2$O/Au. These results form a solid basis for the further development of efficient Cu$_2$O/ZnO solar cells.

Acknowledgments

This work was funded by the California Energy Commission, San Diego State University Foundation grant #55778A/08-02.

References

1 N.A. Mohemmed Shanid, M. Abdul Khadar, *Thin Solid Films* **516** 6245–6252 (2008).
2. G. G. Condorelli, G. Malandrino, and I. Fragala, *Chem. Mater.* **6**, 1861 (1994).
3. Y. Tang, Z. Chen et al., *Materials Letters* **59** (2005).
4. H. Tanaka, T. Shimakawa, T. Miyata, H. Sato and T. Minami, *Appl. Surf. Sci.* **244**, 568 (2005).
5. D.K. Zhang, Y.C. Liu, Y.L. Liu, H. Yang, *Physica B* **351**, 178 (2004).
6. K. Akimoto, S. Ishizuka, M. Yanagita, Y. Nawa, G.K. Paul, T. Sakurai, *Sol. Energ.* **80** 715 (2006).
7. H. Tanaka, T. Shimakawaa, T. Miyata, *Thin Solid Films* **80** 469-470, (2004).
8. A. Mittiga, E. Salza, F. Sarto et al., *Appl. Phy. Lett.* **88**, 163502 (2006).
9. A.M.P. Santos, Edval J.P. Santos, *Materials Letters* **61** 3432–3435 (2007).
10 . H. Chen, J. Zhong, G. Saraf, Y. Lu, D. H. Hill, S. T. Hsu, and Y. Ono, *J. Electron. Mater.* **35**, 1314 (2006).
11.Hanhong Chen, Aurelien Du Pasquier, Gaurav Saraf, Jian Zhong and Yicheng Lu, *Semicond. Sci. Technol.* **23** 045004 (2008).
12 . Hanhong Chen and Yicheng Lu, *Applied Physics Letters* **89** (1), 253513 (2006)

Author Index

Ahn, J.	47	Habuka, H.	81
Ahyi, C.	61	Hirahara, N.	13
Alur, S.	61	Hite, J.	47, 65
Anderson, T.	65	Hobart, K.	65
		Hong, J.	61
Baik, K.	89	Horng, R.	53
Basile, A. F.	95	Hsieh, T.	131
Bellodi, M.	119	Huang, K.	137
Bozack, M.	61	Huang, T.	33
		Huang, T. H.	137
Chang, H.	137		
Chang, L.	33	Jang, S.	89
Chen, C.	33		
Chen, C.	27	Kato, T.	81
Chen, K.	155	Katsumi, Y.	81
Chen, S.	137	Kim, H.	47
Chen, X.	95	Kim, J.	47
Chernyak, L.	3	Kub, F.	65
Chou, M.	33		
Chu, C.	137	Larina, L.	161
		Lee, C.	21, 27, 71
Dai, J.	61	Lee, H.	21
Dai, J.	149	Lim, K.	161
Dashevsky, Z.	3	Lim, W.	89
Dhar, S.	95	Lin, C.	149, 155
Du Pasquier, A.	179	Lin, C.	155
Duan, Z.	179	Lin, M.	149
		Lin, Y.	21
Eddy, Jr., C.	47, 65	Lo, C. C.	131
		Lu, Y.	53
Feldman, L. C.	95	Lu, Y.	179
Flitsyian, E.	3		
Fukae, K.	81	Mastro, M.	47, 65
Furukawa, K.	81	McKittrick, J.	169
		Mishra, K. C.	169
Giroldo Jr., J.	119	Mooney, P. M.	95
Gnanaprakasa, T.	61		
Guo, D.	105, 111	Nakao, M.	13

Onwona-Agyeman, B.	13
Park, M.	61
Pearton, S.	89
Pereira, N.	179
Ren, F.	89
Rozen, J.	95
Sharma, Y.	61
Shatkhin, M.	3
Shevaleevskiy, O.	161
Simonian, A. L.	61
Tadjer, M.	65
Takechi, N.	81
Talbot, J.	169
Tanaka, K.	81
Tao, J. H.	169
Tsvetkov, N.	161
Wang, Y.	61
Wang, Y.	89
Williams, J. R.	95
Wu, M.	137
Wuu, D.	53
Yan, J.	27, 71
Yang, J.	89
Yang, J.	161
Yu, J.	33